BIOCHEMISTRY FOR MATERIALS SCIENCE

AF795081

BIOCHEMISTRY FOR MATERIALS SCIENCE

Catalysts, Complexes and Proteins

AKIO MAKISHIMA

ELSEVIER

Elsevier
Radarweg 29, PO Box 211, 1000 AE Amsterdam, Netherlands
The Boulevard, Langford Lane, Kidlington, Oxford OX5 1GB, United Kingdom
50 Hampshire Street, 5th Floor, Cambridge, MA 02139, United States

© 2019 Elsevier Inc. All rights reserved.

No part of this publication may be reproduced or transmitted in any form or by any means, electronic or mechanical, including photocopying, recording, or any information storage and retrieval system, without permission in writing from the publisher. Details on how to seek permission, further information about the Publisher's permissions policies and our arrangements with organizations such as the Copyright Clearance Center and the Copyright Licensing Agency, can be found at our website: www.elsevier.com/permissions.

This book and the individual contributions contained in it are protected under copyright by the Publisher (other than as may be noted herein).

Notices
Knowledge and best practice in this field are constantly changing. As new research and experience broaden our understanding, changes in research methods, professional practices, or medical treatment may become necessary.

Practitioners and researchers must always rely on their own experience and knowledge in evaluating and using any information, methods, compounds, or experiments described herein. In using such information or methods they should be mindful of their own safety and the safety of others, including parties for whom they have a professional responsibility.

To the fullest extent of the law, neither the Publisher nor the authors, contributors, or editors, assume any liability for any injury and/or damage to persons or property as a matter of products liability, negligence or otherwise, or from any use or operation of any methods, products, instructions, or ideas contained in the material herein.

Library of Congress Cataloging-in-Publication Data
A catalog record for this book is available from the Library of Congress

British Library Cataloguing-in-Publication Data
A catalogue record for this book is available from the British Library

ISBN: 978-0-12-817054-0

For information on all Elsevier publications visit our website at https://www.elsevier.com/books-and-journals

ELSEVIER
Book Aid International
Working together to grow libraries in developing countries
www.elsevier.com • www.bookaid.org

Publisher: Matthew Deans
Acquisition Editor: Christina Gifford
Editorial Project Manager: Andrea Gallego Ortiz
Production Project Manager: Kamesh Ramajogi
Cover Designer: Greg Harris

Typeset by SPi Global, India

CONTENTS

Part Three
Biochemistry Related Materials and Topics

Part Five
Environmental Sciences, Origin and Evolution of Life-Related Molecules and Water on Other Planets, Moons and Dwarf Planets

LIST OF FIGURES

LIST OF TABLES

PREFACE

This book is a collection of omnivorous topics from papers published in 2016–2017 in biochemical sciences and biochemical materials such as drugs, proteins, and catalysts. It explains the up-to-date topics of biochemical sciences using figures and chemical formulas for readers of pure materials sciences, allowing even those who have little knowledge in biochemistry to understand the recent biochemical innovations.

The author realizes that this book may be taken from a shelf by biochemical scientists who have curiosity in up-to-date topics. The author also anticipates that this book may be picked up by pure materials scientists (being industrial or material scientists, inorganic and organic chemists or even cosmochemists) who have an interest in biochemistry.

This book has four purposes. The first is to inform biochemical scientists, who have biochemical backgrounds as professional biochemists, biochemical scientists, pharmacologists, medical doctors, etc., of up-to-date topics in biochemical sciences. Such students or junior and senior scientists can enjoy recent topics from papers published mainly in 2016–2017.

The second purpose is to allow pure materials scientists at all stages (undergraduate and graduate-students, or junior and senior scientists) to understand the essences of recently synthesized or discovered biochemical materials and ideas. All pure materials scientists can enjoy up-to-date innovations in biochemistry and its background. The author hopes that this book will also cultivate literacy in biochemical sciences, allowing pure materials scientists to be able to read and understand relevant journal articles without hesitation. Thus they can enjoy recent topics in biochemical sciences, and those in future as well. For this purpose, this book devoted Part II as an introduction to biochemistry for those who are not familiar with biochemistry.

The third purpose of this book is to be a textbook of biochemical sciences for pure materials science students with their teachers' help. It is also possible for biochemistry students to use this book as a textbook of modern biochemical sciences. The author is concerned that recent students have too narrow a knowledge in sciences. A student in one professional field may know little about other fields in sciences. The author hopes that the views of students in the pure materials sciences (and biochemistry as well) should be widened by using this book. It is clear that the knowledge of other fields can be useful for a reader in progressing within their own main field.

The fourth purpose of this book is for all readers. The readers can enjoy the beauty of chemical formulas of biochemical materials.

The author was once a pure materials scientist in a company, and then became an analytical geochemist, recently publishing two books: *Thermal Ionization Mass Spectrometry (TIMS): Silicate Digestion, Separation and Measurement* (Makishima, 2016), and *Origins of the Earth, Moon, and Life: An Interdisciplinary Approach* (Makishima, 2017). Both books represent the major(s) of the author.

This book is composed of five parts. Part I covers eight topics that the author chose as the most interesting topics in 2016–2017 in biochemistry as well as materials sciences. This part includes important knowledges for general scientists: helium makes compounds; four boron-related topics; inorganic double helix; ultimate vancomycin and tardigrade's survival. In Part II, fundamental knowledges and techniques in biochemistry are shown. This part can be skipped by biochemists. Part III looks at biochemistry-related materials and topics. For example, topics of metallo-porphyrins, new biochemical methods, bioluminescence, new antibiotics, chirality in space and brains, materials and topics on bacteria, plants, animals, proteins, and enzymes are shown. In Part IV, life-sciences related materials and topics are chosen. This part contains human-related topics, such as pain, disease, therapy, drugs, toxic materials, detoxification, teeth, various detectors, nanomachines, click chemistry, and fluorine-containing materials. Part V looks at topics on environmental sciences, origin of life-related molecules, and water on other planets and dwarf planets.

You can start with any chapter that looks interesting, and skip any chapters that you wish. (However, the author suggests that those with little knowledge of biochemistry should read Part II first.) You can glance at any chemical formulas and contemplate the beauty of biochemical materials.

Akio Makishima
Institute for Planetary Materials (IPM), Okayama University, Misasa, Japan

REFERENCES

Makishima, A., 2016. Thermal Ionization Mass Spectrometry (TIMS). Silicate Digestion, Separation, Measurement. Wiley-VCH Verlag GmbH & Co. KGaA, Weinheim, ISBN: 978-3-527-34024-8, p. 348.

Makishima, A., 2017. Origins of the Earth, Moon and Life: A Multidisciplinary Approach. Amsterdam, Holland, Elsevier, Amsterdam, ISBN: 978-0-12-812058-3, p. 266.

PART ONE

The Author’s Choice of 2016–17

CHAPTER ONE

The Author's Choice of 2016–2017

Contents

Biochemistry for Materials Science
https://doi.org/10.1016/B978-0-12-817054-0.00001-1

© 2019 Elsevier Inc.
All rights reserved.

1.1 A STABLE COMPOUND: Na_2He AT HIGH PRESSURE

1.1.1 Diamond Anvil Cell

In order to increase the achievable pressure, the harder material is advantageous. In addition, the massive support theory requires a larger ratio of the bottom of the anvil against the top of the anvil. To satisfy the first criterion, diamond is the best material. In the second criterion, if a 3.5 mm ϕ diamond is pressed with the top of 0.04 mm ϕ, a high ratio of ~7700 times can be obtained.

Fig. 1.1.1A shows a photograph of the two diamonds in the diamond anvil cell. There is a sample at the place indicated by the box. On the right-hand side is the enlarged cross section. The sample is held by the Re gasket. The initial pressure is given by the cell assembly with four small screws (e.g., 6 mm ϕ) as shown in Fig. 1.1.1B, and the white arrow indicates the place of the diamonds. The sample is generally heated by the laser light from the top of the sample through the diamonds to >100 GPa (giga-pascal) and 3000°C.

The largest merit of the diamond anvil cell is that we can see the sample at high temperature and pressure conditions. The disadvantage is that the high pressure area is too small to take out highly pressurized samples. Therefore, the sample is studied in situ using lights (IRs, visible lights, X-rays) through diamonds.

1.1.2 Helium Forms a Stable Compound

Helium is generally understood to be chemically inert. This is due to its extremely stable closed-shell electron configuration, zero electron affinity, and the highest ionization energy of all the elements.

In order to examine this inertness, scientists have searched for helium molecules both theoretically and experimentally. Dong et al. (2017) finally succeeded in discovering a stable helium compound. They used the ab initio evolutionary algorithm USPEX, and scanned helium-sodium interaction over an extremely high pressure range. They then concluded that Na_2He should be thermodynamically stable at the higher pressure of 115 GPa. They subsequently used a diamond anvil cell (described in Section 1.1.1) and succeeded in synthesizing the compound.

Based on the X-ray diffraction, Dong et al. (2017) reported that Na_2He has a fluorite-type structure (see Fig. 1.1.2; sodiums are the spheres, heliums

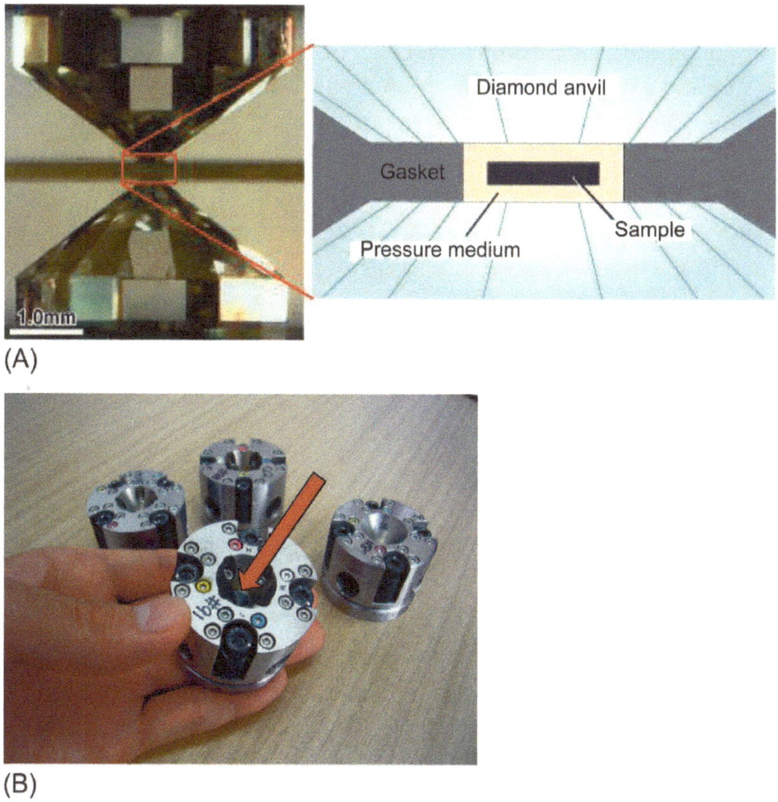

Fig. 1.1.1 The diamond anvil cell. (A) Two diamonds are facing with sandwiching a sample (the box). The right side figure is a schematic diagram of a cross section of the sample, gasket and diamond anvil. (B) The diamond anvil cell assembly. The arrow shows the place of the diamond anvils. (Photographs and figure credits: Dr. Shigehiko Tateno, Earth-Life Science Institute (ELSI), Tokyo Institute of Technology, Japan.)

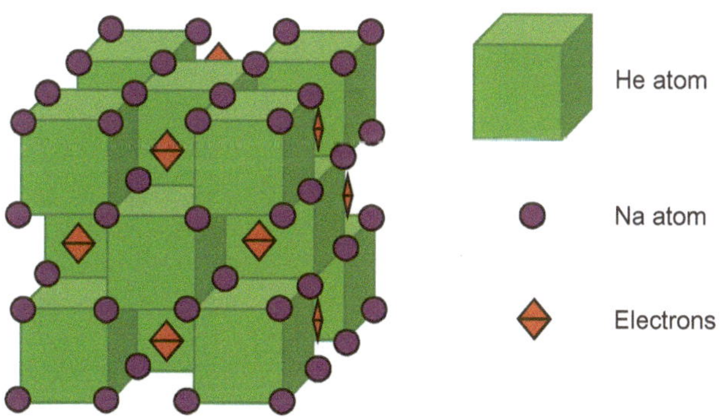

Fig. 1.1.2 The stable structure of Na_2He.

are the cubes, and electrons are localized in the diamond regions), and electrically insulating. The material should remain stable at 1000 GPa at least.

They also found that Na_2HeO should be stable above 15 GPa, but they could not synthesize this yet.

This study requires chemistry textbook sections on noble gases to be rewritten, and shows the importance of high pressure chemistry, because such materials are stable in the interior of the Earth, such as the lower mantle or the core. Thus this topic has been included first in this book.

1.2 BORON CHEMISTRY AND OCTOPUS MOLECULE

1.2.1 Dendrimers and Dendrons

Dendrimers are branched and cascade molecules (see Fig. 1.2.1A). Dendrimers are usually symmetric molecules around the core, and often adopt a spherical three-dimensional morphology. The word "dendron" is often used. Dendrons (see Fig. 1.2.1B) indicate symmetric parts of dendrimers; however, dendrons and dendrimers are sometimes used interchangeably.

Metallodendrimers are metal-containing macromolecules that attract an increased interest in the development of nanometer-scale materials. The incorporation of redox-active units into dendritic structures leads to a wide range of applications such as sensors, catalysts, biomedical materials, and multielectron storage devices.

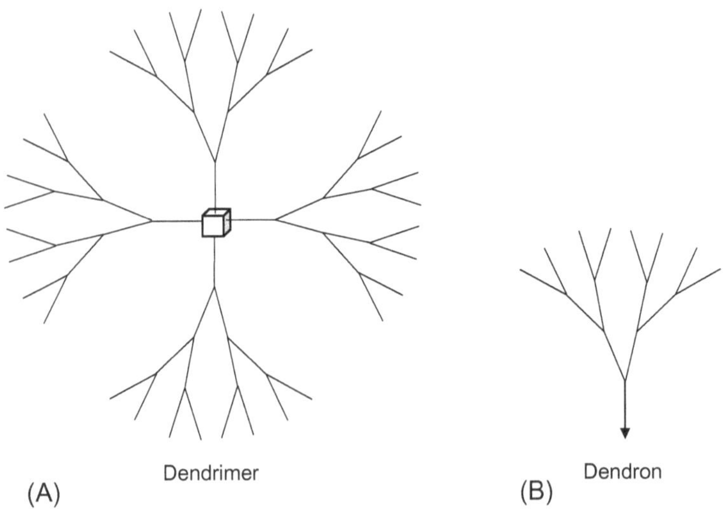

Fig. 1.2.1 Dendrimer (A) and dendron (B).

Dendrimer-type molecules are ideal platforms to construct these devices, which must be designed to satisfy some conditions such processing robust and reversible redox moieties.

1.2.2 Octopus Molecule

The octopus molecule consists of a body and eight arms, and thus looks like an octopus. The body is made of octasilsesquioxane, and its arms are made of long organic molecule. In this section, details for octasilsesquioxane and the long organic molecule (arm) are explained. Cabrea-Gonzalez et al. (2016) synthesized a chemo-octopus molecule, which has so far shown superlative redox powers. In this chapter, the octopus molecule with the octasilsesquioxane as a body and eight metallacarborane-containing dendrons as arms is picked up and explained in detail.

1.2.3 Octasilsesquioxane as a Body

The silsesquioxane is a molecule made of silicons and oxygens with the following chemical formula:

$$\left[RSiO_{3/2}\right]_n \ (n=6, 8, 10, 12, \ldots)\ (R=H, \text{alkyl, aryl, alkoxyl, etc.}) \tag{1.2.1}$$

The silsesquioxane is a colorless solid that forms cage-like or polymeric structures with Si-O-Si linkages and tetrahedral Si vertices. The chemical formulas of Eq. (1.2.1) are sometimes abbreviated as T_8H_8 ($R=H$) or T_8R_8 when $n=8$ (e.g., Cordes et al., 2010).

The octasilsesquioxane is a molecule with $n=8$ in (1.2.1), as shown in Fig. 1.2.2. Polyhedral octasilsesquioxanes (POSSs) are 3D scaffolds that, unlike most dendrimers, are easy to synthesize and exhibit robustness, versatility in functionality, and high solubility. The inorganic core of POSS cages provides mechanical stiffness and thermal stability, whereas the organic pendant groups provide oxidative stability as well as mechanical, optical, or electrical properties. These characteristics make POSSs promising building blocks for hybrid nanomaterials toward technological applications (e.g., Cordes et al., 2010).

For example, POSS compounds are additives in cosmetics. As POSS cages are building blocks for zeolite-like materials, they are used in the wider field of sol-gel processing of polysilsesquioxanes (e.g., Cordes et al., 2010).

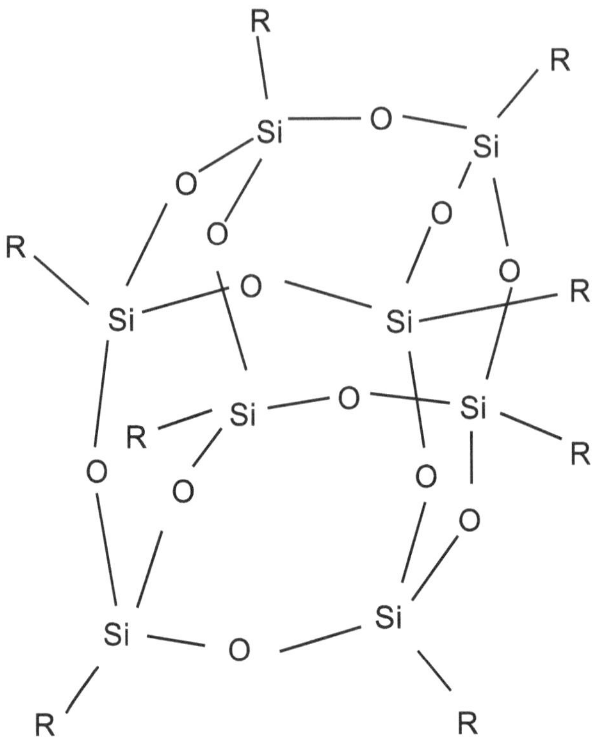

Fig. 1.2.2 Octasilsesquioxane.

1.2.4 Metallocene

A metallocene is a molecule formed by two facing cyclopentadienyl anions (Cp or $C_5H_5^-$) bound to a center metal which is the oxidation state of +2, resulting in the formula $(C_5H_5)_2M$ (see Fig. 1.2.3A). The first metallocene was ferrocene (see Fig. 1.2.3B), which was discovered by Kealy and Pauson (1951) and Miller et al. (1952). The structure of $C_{10}H_{10}Fe$ was determined by Wilkinson et al. (1952) and Fischer and Pfab (1952). Wilkinson and

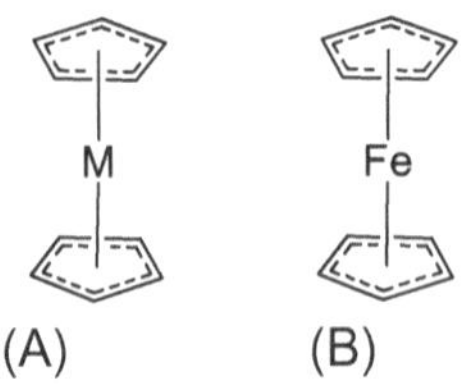

Fig. 1.2.3 Metallocenes: (A) metallocene; (B) ferrocene.

Fischer were awarded the Nobel Prize in Chemistry in 1973 for their work of determining structures of such sandwich compounds.

They found that the cyclopentadienyl ligand (Cp) contributed equally to the bonding that occurred between the d-orbitals of the metal and the p-orbitals of the Cp ligands. This complexing occurs not only for Fe, but also for other transition metals as $[\eta^5\text{-}(C_5H_5)_2M]$ for Co or Ni. The nomenclature of "η^5," which is pronounced "pentahapto," means that five carbon atoms of cyclopentadienyl ring are equally bonded to the metal.

1.2.5 Arms With Electroactive Property

As a particular case, silsesquioxane-cored dendrimers bearing ferrocenyl units were reported to have electroactive properties. However, an important drawback of some ferrocene (Fc; see Fig. 1.2.3B and Section 1.2.4)-functionalized silsesquioxane dendrimers is lack of solubility in common solvents and self-assembling properties Bauduin et al., 2011; Brusselle et al., 2013; Farràs et al., 2012).

To overcome this drawback, the chemo-octopus used a metallocarborane as an arm (see Fig. 1.2.4; the metallocarborane is explained in Section 1.2.7). Consisting of eight arms and a silsesquioxane body, the chemo-octopus was synthesized (Cabrea-Gonzalez et al., 2016).

The chemo-octopus, which is shown in Fig. 1.2.5, shows not only high thermal and chemical stability (Sivaev and Bregadze, 1999), but also high solubility in organic solvents. Furthermore, the molecule acts as an independent carrier of eight electrons essentially at the same potential. The molecule that carries, stores, and transfers many uniform electrons is important in developing molecule-based electronics.

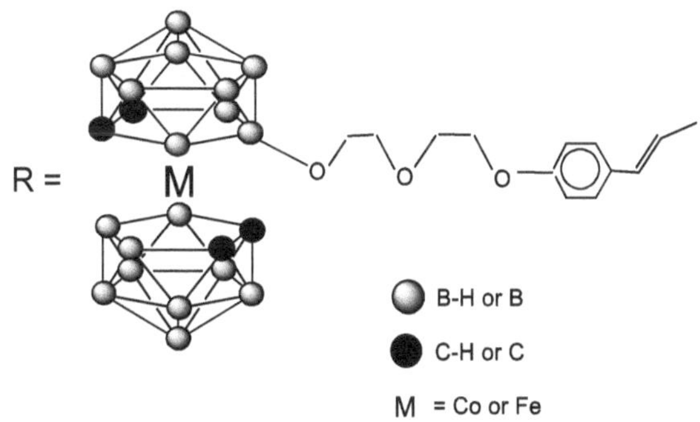

Fig. 1.2.4 An arm (R) of chemo-octopus.

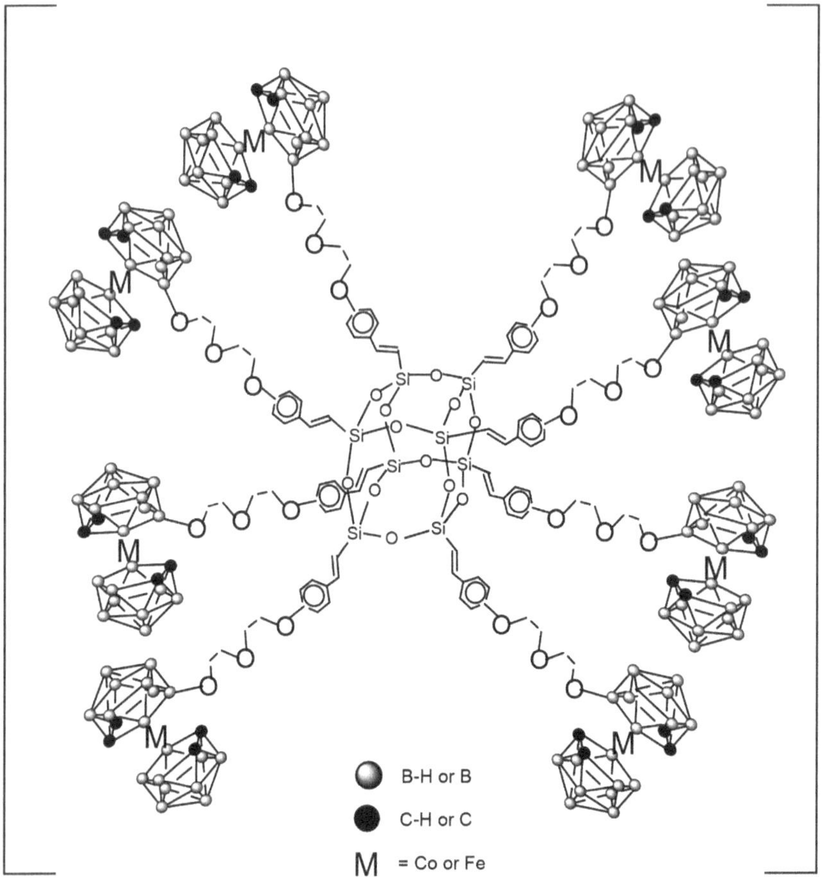

Fig. 1.2.5 The chemo-octopus.

1.2.6 Boranes

Boranes are compounds with boron (B) and hydrogen (H). From 1912 to 1936, Alfred Stock and coworkers synthesized and studied many boranes, for example, B_2H_6 (diborane (6); b.p. in vacuum −92.59°C; see Fig. 1.2.6), B_4H_{10} (tetraborane (10); 18°C; Fig. 1.2.7A), B_5H_9 (pentaborane (9); 60°C; Fig. 1.2.7B), B_5H_{11} (pentaborane(11); 65°C; Fig. 1.2.7C), B_6H_{10} (hexaborane(10); 108°C; Fig. 1.2.7D), $B_{10}H_{14}$ (decaborane(14); 213°C; Fig. 1.2.7E), etc.

The number in parentheses after the borane indicates the number of hydrogen atoms in one molecule. Diborane(6) (B_2H_6) is flammable gas in

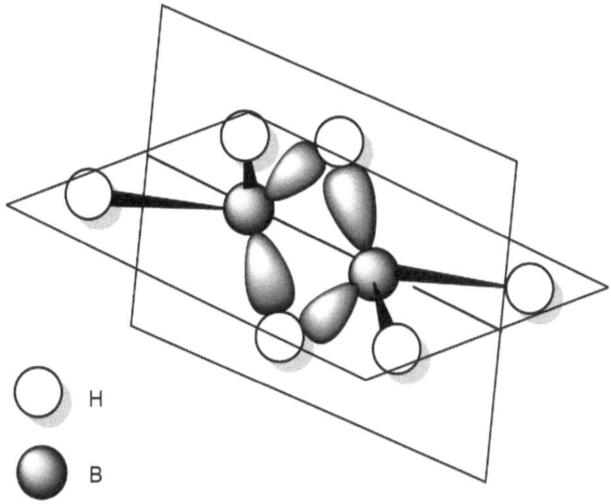

Fig. 1.2.6 Diborane (6) (B_2H_6).

air. Thus, in order to deal with these boranes, Stock developed vacuum experimental techniques. In addition, boranes have interesting chemical properties as well as remarkably interesting structures. For example, the structure of diborane (6) is shown in Fig. 1.2.6. Along a B-B line, there are two perpendicular planes. On one plane there are four usual sp^3 B-H bondings; however, on the other perpendicular plane, two prototypal three-center two-electron (3c-2e) hydrogen bridge bondings are formed. One hydrogen atom is bonded to two boron atoms, each of which donate one electron, so that two boron atoms give two electrons to two hydrogen atoms to form two B-H, 3c-2e bondings.

1.2.7 Metallocarboranes

Carboranes (more formally but less commonly called carbaboranes) are polyhedral boranes that contain carbon atom(s) in the framework of borons. The C_nB_m skeleton is numbered starting with an apex atom; successive rings (belts) of atoms are then numbered in a clockwise direction. The C (or other) atoms are given the lowest numbers possible in this scheme.

As a cyclopentadienyl ion ($C_5H_5^-$) (see Fig. 1.2.8A), the 1,2-$C_2B_9H_{11}^{2-}$ ion (see Fig. 1.2.8B), commonly called dicarbollide ions, are capable of coordinating with metal ions. This ion was used in the arm of the chemo-octopus (see Fig. 1.2.4).

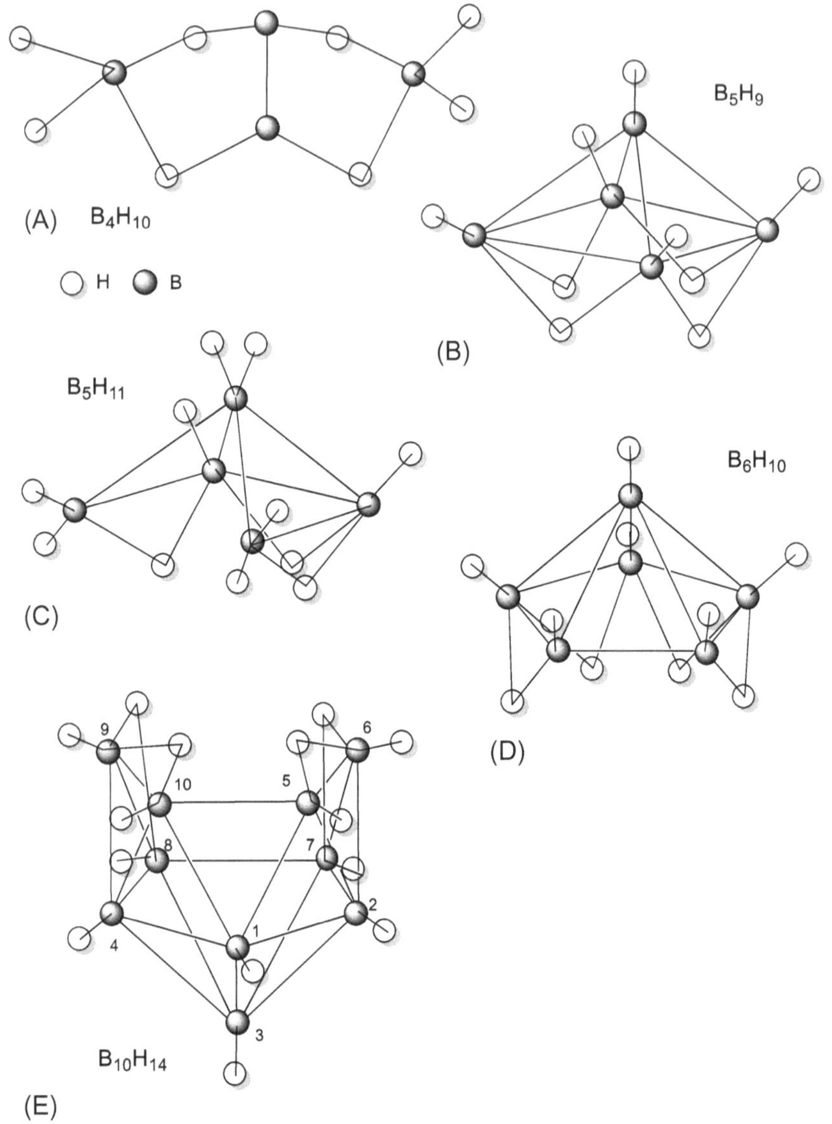

Fig. 1.2.7 Boranes: (A) tetraborane (10); (B) pentaborane (9); (C) pentaborane (11); (D) hexaborane (10); (E) decaborane (14).

1.2.8 NMR Spectroscopy

Nuclear magnetic resonance (NMR) is a phenomenon where nuclei in a magnetic field absorb and re-emit electromagnetic radiation. This resonance energy (***E***) or frequency (***ν***) is determined by the strength of the magnetic field (***B***), the magnetic properties of the isotope of the atom (***μ***):

$$\boldsymbol{E} = -\boldsymbol{\mu} \cdot \boldsymbol{B} \tag{1.2.2}$$

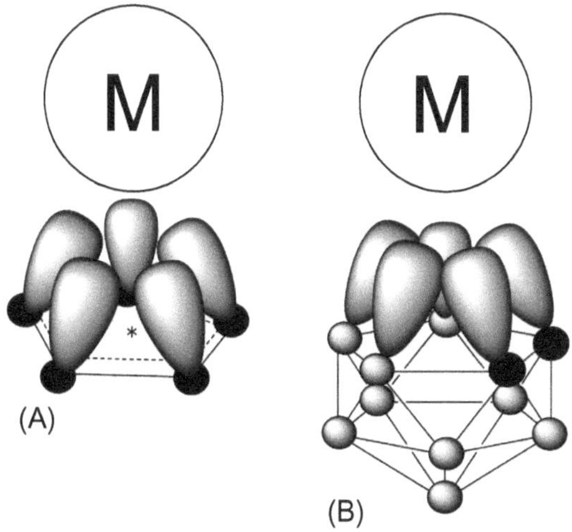

Fig. 1.2.8 (A) Cyclopentadienyl (C_5H_5-) and (B) 1,2-$C_2B_9H_{11}$ 2- ions coordinating metal ions (M = Fe^{2+} or Co^{2+} ions). The figure shows only half of the metal-complex. Black and white balls indicate carbon and boron atoms, respectively.

and

$$\mu = \gamma m \hbar \tag{1.2.3}$$

where γ, m, and $\hbar$ are the z-component of the magnetic moment, the spin status (e.g., $\pm 1/2$), and the reduced Planck constant, respectively. In NMR, the splitting energy ($\Delta \boldsymbol{E}$) between $+m$ and $-m$ is determined at the constant magnetic field ($\boldsymbol{B}_0$):

$$\Delta \boldsymbol{E} = 2\gamma m \hbar \boldsymbol{B}_0 \tag{1.2.4}$$

From Eq. (1.2.4), it can be seen that nuclear magnetic resonance does not occur for the isotope of the spin status of zero. All nuclei with even protons and neutrons have no nuclear spin; therefore, the spin status becomes zero, and the nuclear magnetic resonance is not observed. The resonance energy changes according to the magnetic moment (γ) around the target nuclei at the constant magnetic field.

The resonance frequency is similar to VHF and UHF television broadcasts (60–1000 MHz). NMR allows the observation of specific quantum mechanical magnetic properties of the atomic nucleus. The magnetic property of the nucleus changes according to the shielding effects by the electrons around the target nucleus, because the electron shielding reduces the magnetic field around the nucleus.

Therefore, NMR spectroscopy is a principal technique to obtain the topology, dynamics, and three-dimensional information of molecules. The proton-NMR is also used routinely in medical imaging techniques, as magnetic resonance imaging (MRI). The word "nuclear" may remind nonscientists of "atomic bombs or atomic reactions," and use of this word was therefore avoided in MRI.

Commonly used nuclei for NMR analyses are ^{1}H (spin −1/2), ^{11}B (spin 3/2), ^{13}C (spin −1/2), ^{14}N (spin −1), ^{19}F (spin −1/2), ^{31}P (spin −1/2), etc. Therefore, boranes and carboranes are suitable for application of NMR, resulting in rapid advances of their structural studies.

1.2.9 ^{11}B NMR

The nuclei of ^{11}B and ^{10}B are made of five protons, and six and five neutrons, and have the nuclear spins of 3/2 and 3, respectively. Therefore both ^{11}B and ^{10}B show nuclear magnetic resonance; however, ^{11}B shows sharper peaks than ^{10}B, and therefore it is used in boron-NMR.

Fig. 1.2.9 shows some general features of ^{11}B NMR spectra for ^{11}B-^{1}H-abundant boron-hydride materials. When only one H atom is attached to B (≡B—H), the NMR spectrum shows the single doublet peaks (Fig. 1.2.9A). When two H atoms are connected with B (=BH_2), the B NMR spectrum becomes a triplet peaks (Fig. 1.2.9B). As one of the doublet peaks overlaps together, the area of the overlapped center peak becomes twice as large as that of one of two side peaks. When three H atoms are bound to B (-BH_3), the NMR spectrum becomes symmetrical quartet peaks (Fig. 1.2.9C). The area of one of two side peaks is one-third that of one of the doublet peaks at the center.

Fig. 1.2.10 shows schematically the actual ^{11}B NMR spectrum of $B_{10}H_{14}$ (see Fig. 1.2.7E) at 64 MHz. The vertical axis is the intensity of the NMR signal, and the horizontal axis is the intensity of the magnetic field. The number in the figure indicates the boron atom numbered in Fig. 1.2.7E. The peak area (or rough peak intensity) corresponds to the number of boron atoms. Therefore, the peak area of A is equal to that of B, and the peak area of C is half of those of A or B.

The ^{11}B NMR technique combined with ^{1}H NMR has been a very powerful method for structural studies of boranes.

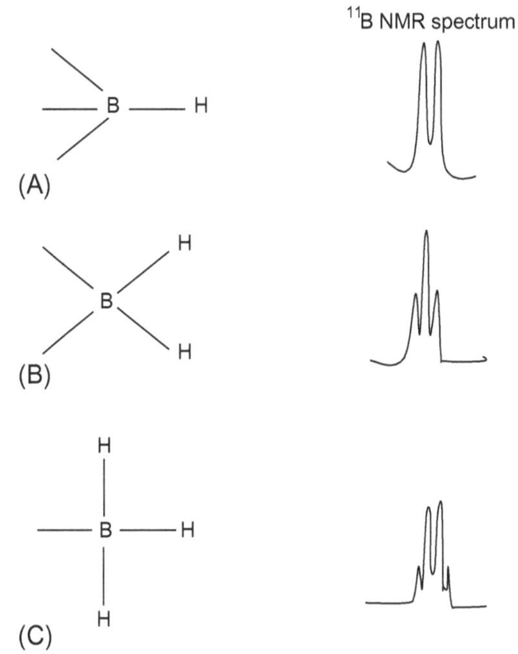

Fig. 1.2.9 Some general features of ^{11}B NMR spectra. (A) When only single H atom is attached to B. (B) When two H atoms are connected to B. (C) When three H atoms are bound to B.

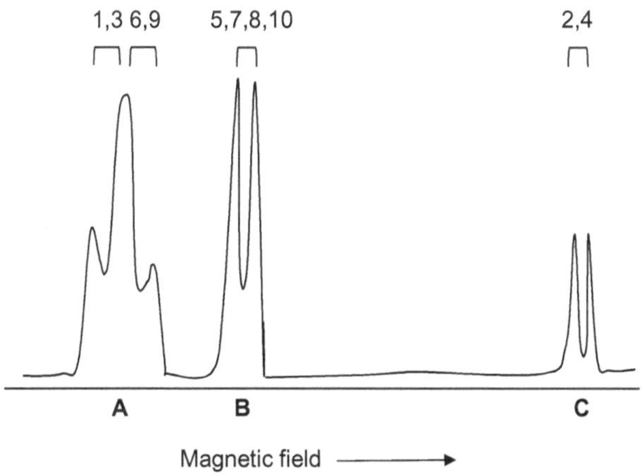

Fig. 1.2.10 The schematic ^{11}B NMR spectrum of $B_{10}H_{14}$. The number in the figure corresponds to the number of the boron atom in Fig. 1.2.7E.

1.3 SYNTHESIS OF THE HETEROCYCLE MISSING LINK: B_3NO_2

Chemists have been trying to synthesize various kinds of six-membered heterocycles containing boron, carbon, nitrogen, and oxygen (see Fig. 1.3.1) to obtain new opto-electronic devices or organo-catalysts. As shown in Fig. 1.3.1, carbon atoms of benzene (C_6H_6) are substituted by one, two, or three boron atoms, substituting another carbon atom into nitrogen or oxygen atoms. When the benzene ring contains three boron atoms with three nitrogen atoms, it is a borazine (B_3N_3) ring; this was synthesized by Stock and Pohland (1926). When the benzene ring is substituted by three boron atoms and three oxygen atoms, it is a boroxine (B_3O_3) ring (Brown, 1972). However, the benzene ring with three boron atoms, two oxygen atoms, and one nitrogen atom (B_3NO_2; 1,3-dioxa-5-aza-2,4,6-triborinane; DATB) has not been synthesized, and is called the "missing link" between borazine and boroxine rings.

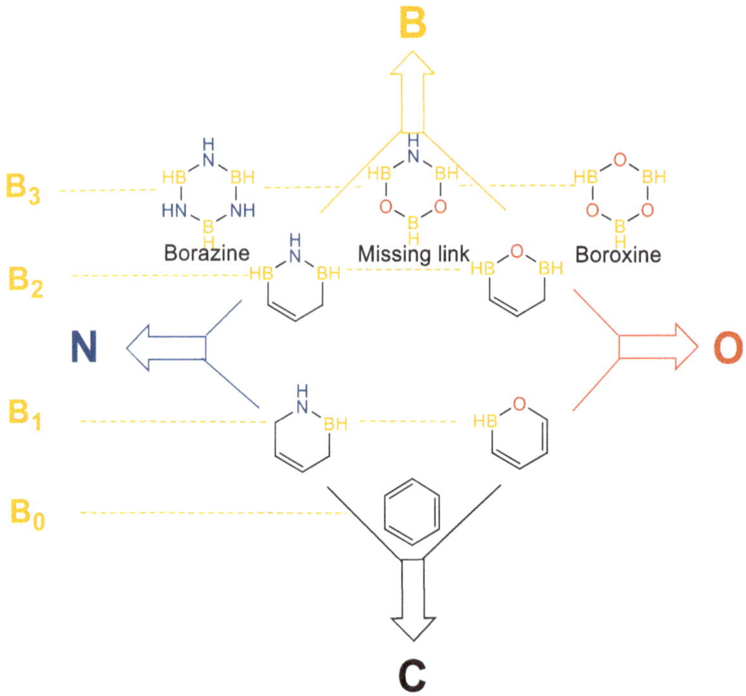

Fig. 1.3.1 A schematic diagram of the B—O—C—N square and a position of a B_3NO_2 ring system called the "missing link." The vertical numbers, B0, B1, B2, and B3, indicate the number of boron atoms in the ring.

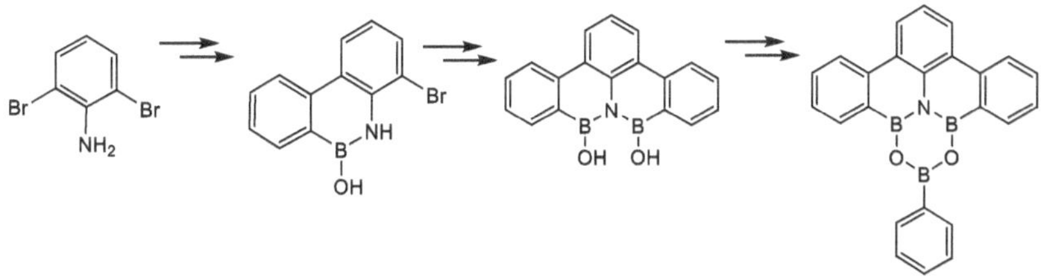

Fig. 1.3.2 Synthesis of a B_3NO_2, ring motif, 1,3-dioxa-5-aza-2,4,6-triborinane (DATB).

Noda et al. (2017) synthesized the B_3NO_2 (DATB) ring motif for the first time. The m-terphenyl-templated DATB derivatives were stepwise synthesized starting from bromine-substituted aniline (see Fig. 1.3.2). Noda et al. realized that the multiple Lewis acidic boron atoms in the DATBs should make the new molecules good catalysts. For example, they succeeded in the direct amidation of carboxylic acids with amines. DATB derivatives with bulkier substituted phenyl groups containing a hydroxyl azaborine give high yields for a broad range of amidations.

The author sees not only the beauty of chemistry in this motif, but also possibilities of new applications.

1.4 BORON NITRIDE (BN) AS CATALYST FOR OXIDATIVE DEHYDROGENATION OF PROPANE

1.4.1 Introduction

The oxidative dehydrogenation of propane (ODHP) has been a challenging reaction:

$2CH_3\text{-}CH_2\text{-}CH_3 + O_2 \rightarrow 2\ CH_2{=}CH\text{-}CH_3 + 2\ H_2O$

While propene is produced by steam cracking of large hydrocarbons in naphtha, the recent trend to use shale gas greatly increases the availability of low-cost ethane, but also results in a gap between demand and supply of propene, as well as other higher olefins.

Although nonoxidative propane dehydrogenation is the technology used today, ODHP has the potential to improve reaction efficiency over non-oxidative one, because themodynamics favors the exothermic former one at lower reaction temperatures. Thus the life of the catalyst could be longer due to prevention of coke deposition on the catalyst's surface. It is estimated that the potential energy savings for transferring to ODHP would be ~45% of the energy consumption. This gain in reaction efficiency, together with the fact that demand for propene is ~100,000 kt, is significant for the potential impact of ODHP if it could be done on a large scale. The key scientific challenge that must be overcome is the prevention of the facile over-oxidation into more thermodynamically stable CO or CO_2.

At the time of writing, supported vanadium oxide catalysts (V/SiO_2) show the most promising activity for ODHP, owing to the favorable redox properties of the active vanadium sites. However, despite decades of research, propene selectivity remains low, even at moderate propane conversion. At 10% propane conversion, for example, selectivity of the propene drops to <60%.

1.4.2 Properties of Boron Nitride

Boron nitride is expressed as BN, which is a heat and chemically refractory material. The hexagonal form (h-BN) as graphite is the most stable and soft of polymorphs, which is used in lubricants and as an additive in cosmetics. Cubic BN (c-BN) is analogous to diamond, because it is hard, highly thermally, and chemically stable, but only slightly softer than diamond. The rare wurtzite form (w-BN) has the same structure as lonsdaleite, which has the hexagonal polymorph of carbon.

1.4.3 Boron Nitride as Catalyst

Grant et al. (2016) found both hexagonal boron nitride (h-BN) and boron nitride nanotubes (BNNTs) can work as a catalyst for ODHP. BN materials are the "inorganic analogues" of graphene, which shows the catalyst for alkane oxidation. BN was directly compared with the V/SiO_2 catalyst. These materials were put into a quartz tube reactor heated to 460–500°C under flowing propane, oxygen, and nitrogen as an inert carrier gas. Reaction parameters such as temperature, catalyst mass, total gas flow-rate, and partial pressures of propane ($P_{C_3H_8}$) and oxygen (P_{O_2}) were then varied.

BN materials result in extraordinary selectivity to propene. The h-BN afforded 79% selectivity to propene at 14% propane conversion, while, V/SiO_2 gave 61% propene selectivity at only 9% propane conversion.

When using V/SiO_2, the main byproducts are CO*x* (CO and CO_2) of 33% at total selectivity of the 9% propane conversion. In contrast, the main byproduct of the BN materials was ethene. Both ethene and propene selectivities were 14% at 80% propene conversion, respectively, using h-BN. BN materials remain stable for >32 h.

1.4.4 Mechanism of Catalytic Reaction

The similar product distributions for both h-BN and BNNTs suggested a similar reaction mechanism for these BN materials. However, BNNTs exhibit a ten-fold higher rate of propane conversion than h-BN, which partly reflects the higher surface area of BNNTs than that of h-BN.

An oxygen molecule bonded to one B atom and one N atom (>B-O-O-N<) acts as active site for the ODHP reaction (see Fig. 1.4.1A).

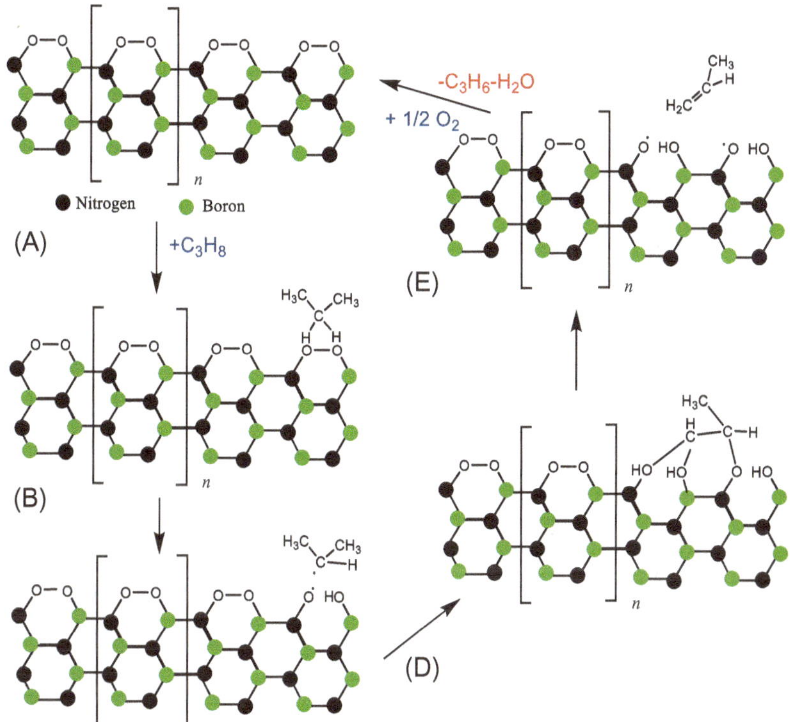

$$2C_3H_8 + O_2 \longrightarrow 2C_3H_6 + 2H_2O$$

Fig. 1.4.1 Boron nitride (BN) catalyst converts propane to propene. (A) Oxygen molecules are bound at the edge of the BN crystal. (B) When propane molecules are supplied, hydrogen atoms of the center carbon are attached on the oxygen molecules. One hydrogen is bounded on the boron oxygen, and the other hydrogen is on the nitrogen oxygen on the >B—O—O—N< site. (C) The O—O bond breaks. The B—OH, N—O• (• means a radical) and •CH(CH3) are formed. (D) Two hydrogens of one of carbon of the propane react with other side of >B—O—O—N< site. (E) The O—O bond of the other side breaks, and the two B—OH, two N—O• and •CH_2—•$CH(CH_3)$ form. The last molecule becomes propene (CH_2=$CH(CH_3)$). As the oxygen is supplied, the radicals on the two >B—OH •O—N< sites return to the two >B—O—O—N< sites and H_2O.

The >B-O-O-N< site can be considered as an inorganic peroxide species, which can perform oxidation reactions. At the >B-O-O-N< site on which propane is attached (Fig. 1.4.1B), the O—O bond is broken, and a BOH species and one nitroxyl-radical are formed (Fig. 1.4.1C). The secondary-propyl radical then reacts rapidly with the nitroxyl-radical, forming a propyl intermediate (>N-O-$CH(CH_3)_2$) (see Fig. 1.4.1C). Then two hydrogens of the carbon of the propane react with the aside >B-O-O-N< site, breaking the O—O bond (see Fig. 1.4.1D). The broken O—O bond forms the >B-OH, >N-O• and •CH_2-•$CH(CH_3)$. The last molecule becomes propene ($CH_2=CH(CH_3)$). As the oxygen gas is supplied, the radicals on the >B-OH •O-N< sites return to the >B-O-O-N< site, forming H_2O (Fig. 1.4.1A and E).

Nobody has imaged that BN becomes a good catalyst for any reactions. This ODHP reaction with the BN catalyst has an impact on the petrochemical industry.

1.5 DECARBOXYLATIVE BORYLATION

1.5.1 Introduction

Boronic acids and their esters are becoming more and more important in chemistry. The boronic acids have been applied not only in the famous Suzuki coupling reaction (Suzuki, the Nobel Prize lecture in 2011; see Section 1.5.2), but also in numerous applications such as material sciences (Brooks and Sumerlin, 2016) or drug syntheses (Trippier and McGuigan, 2010; Draganov et al., 2014).

Boronic acids have Lewis acidity, reversible tendency to reacts with various nucleophiles (e.g., alcohols and amines), and the ability to form hydrogen bonds. Therefore, boronic acids have unique chemical and biological properties. For example, in medical science, boronic acids have been related to bioisosteres. The bioisosteres mean replacements of structural motifs of similar physical and chemical properties, such as carboxylic acids (Ballatore et al., 2013), to alter the physiochemical properties of leading candidates. Alkyl boronic acids and α-amino boronic acids are addressed as potent protease inhibitors.

Recently, two alkyl boronic acids were approved by the FDA for various oncology indications: Ninlaro and Velcade (see Fig. 1.5.1). However, these promising medicinal scaffolds are not fully utilized because of the difficulty of their preparation. Only a few alkyl boronic acids have been synthesized for biological evaluation.

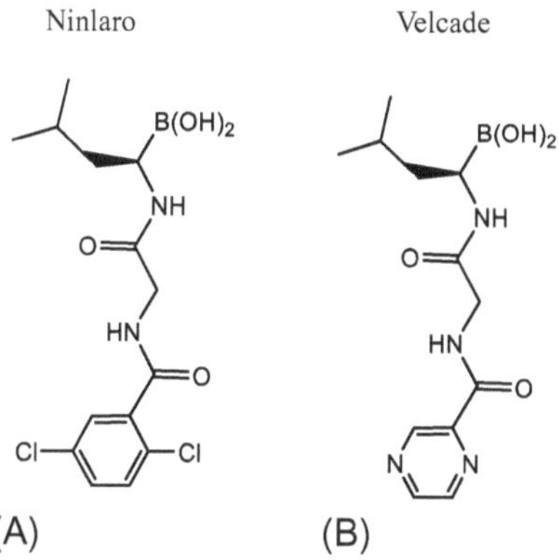

Fig. 1.5.1 (A) Ninlaro (Ixazomib) and (B) Velcade (Bortezomib).

Although carboxylic acids are everywhere in nature and inexpensive, most boronic acids are synthesized. The retrosynthetic analysis using alkyl boronic acids is used to search for drug candidates. There are various methods to make alkyl C—B bonds, such as hydroboration of alkenes, transmetalation with alkyl organolithium species, or conjugate addition. Although these approaches are efficient, there are few methods that use naturally existing or readily available starting materials. For example, the synthesis of an engineered amino acid (AA) is required. In order to test systematically the relationship of leading compounds between the structure and the activity, each analog must be synthesized one by one.

If the direct transformation of the carboxylic acid containing native peptide into the corresponding boronic acid is possible, an easier and more logical approach would be available.

To overcome this problems, Li et al. (2017) have invented a simple method using nickel-catalyzed decarboxylative borylation. Many boronates that required lengthy synthesis have been obtained directly from the corresponding carboxylic acids. Furthermore, Li et al. (2017) found three useful elastase inhibitors of small molecule because the new method can transform native peptides into α-amino boronic acids (not shown).

1.5.2 Suzuki Coupling Reaction

In chemistry, carbon-carbon bond formation is a key step for the synthesis of organic materials. The (sp^2)C-B compounds (aryl- and 1-alkenyl boron derivatives) and (sp^3)C-B compounds (alkylboron compounds) readily

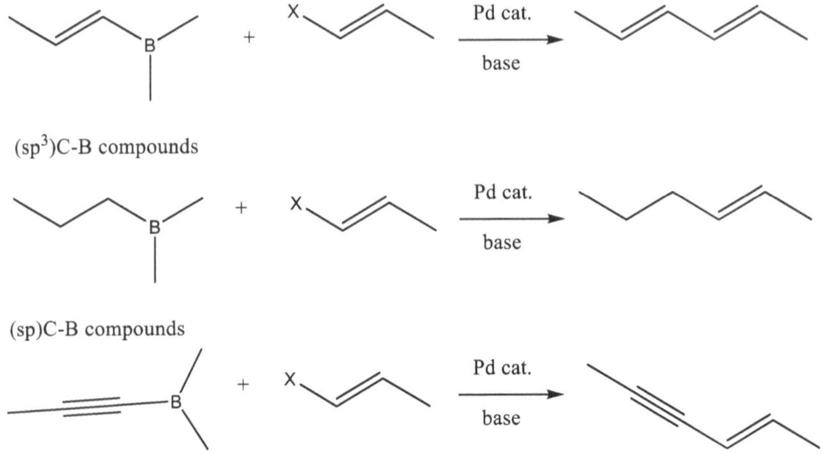

Fig. 1.5.2 Examples of Suzuki coupling reactions.

cross-couple with organic electrophiles to give coupled products in high yield (see Fig. 1.5.2). The (sp)C-B compounds (1-alkynylboron derivatives) also react with organic electrophiles to produce the cross-coupled products (see Fig.1.5.2).

1.5.3 Details of the Decarboxylative Borylation Reaction

To summarize the method of Li et al. (2017), first of all, the carboxylic acids (Fig. 1.5.3A) are converted into the redox-active esters (RAEs) (Fig. 1.5.3B). Next, the RAEs are converted into boronate esters (Fig. 1.5.3C) in three stages: (i) the preparation of catalyst mixture; (ii) the preparation of $[B_2pin_2Me]Li$

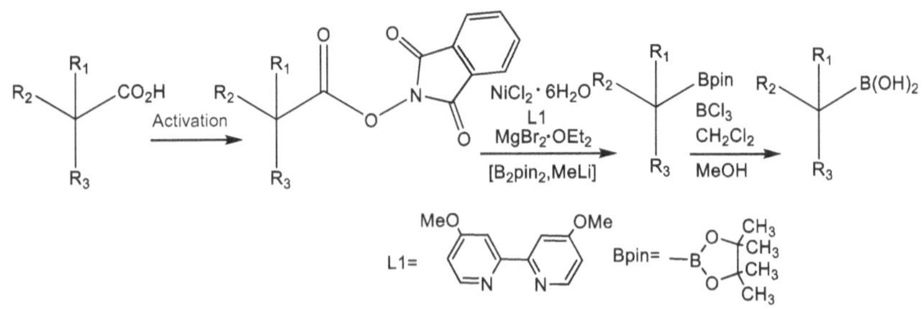

(A) (B) (C) (D)

Fig. 1.5.3 Decarboxylative borylation reaction. (A) The starting materials (the carboxylic acids). (B) The redox-active esters (RAEs). (C) Boronate esters. (D) Boronic acids.

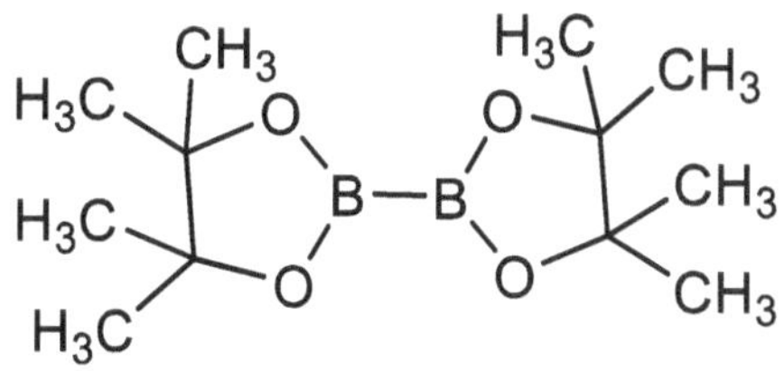

Fig. 1.5.4 Bis(pinacolato)diboron; B_2p_2.

complex (B_2pin_2 = Bis(pinacolato)diboron, see Fig. 1.5.4); and (iii) the nickel catalyzed decarboxylative borylation reaction. Finally, the boronate ester is decomposed into the boronic acid (Fig. 1.5.3D).

The reaction developed by Li et al. (2017) is started by the preparation of the redox-active esters (RAEs, e.g., N-hydroxyphthalimide ester abbreviated as NHPI ester, Fig. 1.5.3B) derived from alkyl carboxylic acids (Fig. 1.5.3A) as convenient agents for alkyl halides in nickel or iron catalyzed cross-coupling reactions.

These useful intermediates, most commonly used in amide bond forming reactions, have enabled practical methods of C—C bond formation in various modalities, including decarboxylative Negishi (e.g., Qin et al., 2016), Suzuki (Wang et al., 2016), and Kumada (Toriyama et al., 2016) couplings.

NHPI ester (Fig. 1.5.3B) proved to be the best substrate for borylation with B_2pin_2. The inexpensive combination of $NiCl_2 \cdot 6H_2O$ and bipyridine ligand (L1) was the best catalyst system after an extensive screening. The choice of solvent was also critical. A binary mixture of tetrahydrofuran (THF) and dimethylformamide (DMF) was the best mixture. Pre-mixing methyl lithium with B_2pin_2 was necessary to activate the diboron species for transmetalation.

A pinacol boronate ester product (Fig. 1.5.3C) was directly obtained from the carboxylic acid in comparable yields using a one-pot procedure where RAE (Fig. 1.5.3B) was formed in situ. Overall, the reaction proceeded smoothly for 2h (1h at 0°C and 1h at room temperature from Figs. 1.5.3B to C and C to D, respectively).

1.5.4 Applications

Li et al. (2017) explored the applicability of decarboxylative borylation reaction. After redox-active esters were prepared from a broad selection of primary, secondary, and tertiary carboxylic acids, the decarboxylative borylation reaction proceeded without any issues. These selections, which contain acyclic, cyclic, caged, bridgehead, fluoroalkyl, and benzylic acids, were transformed to the corresponding pinacol boronate esters (Bpin ester) smoothly.

1.6 INORGANIC DOUBLE HELIX

1.6.1 Introduction

A double helix is a structure that attracts attention from not just chemists but most scientists. The double helix was first found in deoxyribonucleic acid (DNA) by Watson and Crick (1953), and many chemists tried to make similar materials. A helical structure was found in polytetrafluoroethylene (PTFE) at $<20°C$ (Bunn and Howells, 1954). The double helices are building blocks often found in living systems and carbon nanotubes (Fujimori et al., 2013).

However, the double helices have not yet been found in the inorganic chemistry world where no carbons are used. It is expected that the inorganic double helical materials should have excellent mechanical, electrical, and optical properties (Zhao et al., 2014; Pfister et al., 2016). The existence of a simple atomic chain was predicted for LiP, but the double helix of LiP has not been realized yet (Ivanov et al., 2012). LiP reported in the 1980s contained only a P-helix coordinated by lithium.

Several experiments have been attempted to make inorganic double helical materials. Double helical selenium was observed in traces using double-walled carbon nanotubes as a mandril. From a nanometer to micrometer scale, helical arrangements of semiconductors such as CdTe/CdS (Tang et al., 2006; Srivastava et al., 2010), or metals as gold using DNA as a mold (Sharma et al., 2009) have been achieved.

1.6.2 A Double Helical Vanadium-Oxo Material

A more complex double helical material is $[(CH_3)_2NH_2]K_4[V_{10}O_{10}(H_2O)_2(OH)_4(PO_4)_7]\cdot 4H_2O$ (**1**), which is a hydrothermally synthesized vanadium-oxo material (Soghomonian et al., 1993; see Fig. 1.6.1.). The building block shown in Fig. 1.6.1 forms chiral double helices (see Fig. 1.6.2), which contains interpenetrating spirals of vanadium oxo pentamers bonded together by P^{5+}. These double helices intertwine with each other, generating tunnels and cavities that are filled with $(CH_3)_2NH_2^+$ and K^+ cations, respectively.

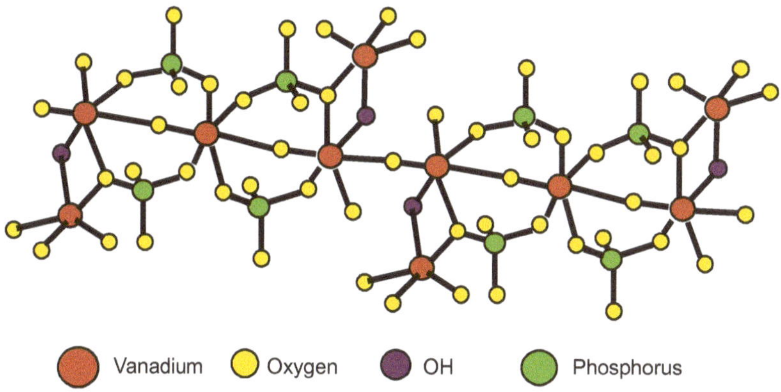

Fig. 1.6.1 One crystallographically independent pentameric vanadium oxo building blocks in the $[(CH_3)_2NH_2]K_4[V_{10}O_{10}(H_2O)_2(OH)_4(PO_4)_7]{\cdot}4H_2O$ (**1**) after Soghomonian et al. (1993).

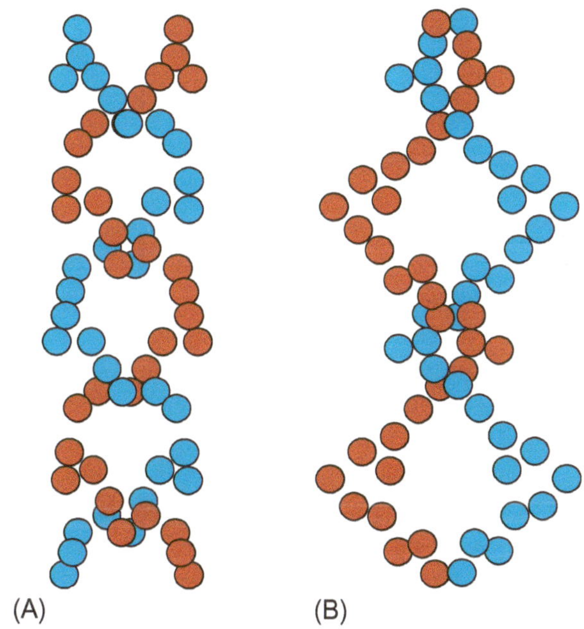

Fig. 1.6.2 Two vanadium oxo spirals that interpenetrate to form the double helices. Red (dark gray in print versions) and blue (light gray in print versions) balls indicate vanadium oxos, respectively. (A) View down $[1\bar{1}0]$. (B) View down [100].

Hydrothermal treatment of KVO_3, V, H_3PO_4, $CH_3PO(OH)_2$, $(CH_3)_2NH$, and H_2O in a mole ratio of 2.35:1:10.18:3.31:8:1137 for 4 days at 200°C gives >85% yield of monophasic, dark-blue tetragonal bipyramids of (**1**). It is surprising not only that the complex structure of (**1**) is self-assembled from such structurally simple starting materials, but also that it is formed in quantitative yield. These optimized synthetic conditions were determined empirically after the observation of (**1**) in other reactions with the same starting materials but with different mole ratios. Although the role of the methylphosphonic acid is not known and it is not incorporated in the product, it is crucial to obtain the high yield of (**1**) under these conditions (Soghomonian et al., 1993).

1.6.3 A Double Helical SnIP

Pfister et al. (2016) prepared bulk SnIP phase on a gram scale in a fast and simple solid state annealing reaction from tin, phosphorus, and tin tetraiodide. In order to systemize the description of SnIP, Pfister et al. (2016) discuss the determination of bulk and nanosized properties separately. SnIP crystallizes monoclinically, in space group $P2/c$ (No. 13) containing $Z = 14$ units in the unit cell and lattice parameters of $a = 7.934$ Å, $b = 9.802$ Å, $c = 18.439$ Å, and $\beta = 110.06$ degrees (see Fig. 1.6.3A). Two substructures of different elemental composition, a helical chain of phosphorus and a second helical chain of SnI, form SnIP double helices of 0.98 nm diameter.

The single rods are either left (M-helix) or right-handed (P-helix), and thus have opposite chirality (see Fig. 1.6.3A). Both sets of double helices are assembled in a pseudo-hexagonal arrangement of rods along the a-axis. Seven SnIP units per two chain windings are present in each rod, resulting in a 7/2 helix (see Fig. 1.6.3B).

Raman spectra show vibration modes of P—P, Sn—P, and Sn—I bonds. SnIP starts to decompose at >600 K by evaporation of SnI_4 and P_4. Pfister et al. (2016) calculated the band gap of 2.28 eV for a single SnIP nanorod. The double helical materials are expected to be a special semiconductor, and this is just the beginning of the research.

1.7 ULTIMATE VANCOMYCIN BY MODIFICATION

Vancomycin (Vancocin) (Fig.1.7.1), which was a natural product of bacteria, has been an antibiotic of the last method, because it is the only effective antibiotic against antibiotic-resistant bacteria after all other drugs

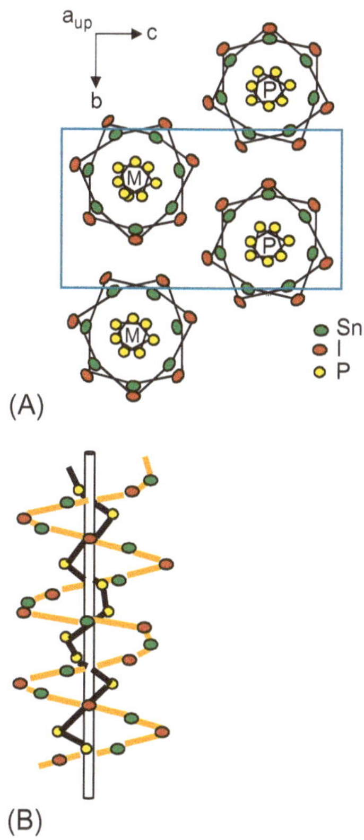

Fig. 1.6.3 A double helical SnIP. (A) Crystal structure projected from *a*-axis. The [SnI] and [P] helices form a double helical SnIP strand. The box shows the unit shell. M and P helices show left- and right-handed ones. (B) Schematic diagram of the double helix arrangement. The [SnI] and [P] helices rotate along the hypothetical axis shown as the white rod. The captions are the same with (A).

did not work. However, some bacteria have evolved to be defensive against vancomycin; thus, there are no antibiotics against them.

Recently, Okano et al. (2017) modified vancomycin and improved antimicrobial potency. They made three types of modifications: (i) a methylene group substituted for an amide carbonyl (#4); (ii) C-terminal peripheral modification, introducing a quaternary ammonium salt (#13); and (iii) (4-chlorobiphenyl)methyl (CBP) addition to the vancomycin disaccharide (#18), of which the numbers in parentheses correspond to the compounds in Fig. 1.7.1, respectively. The antimicrobial potency of each compound is indicated by the minimum inhibitory concentration MIC

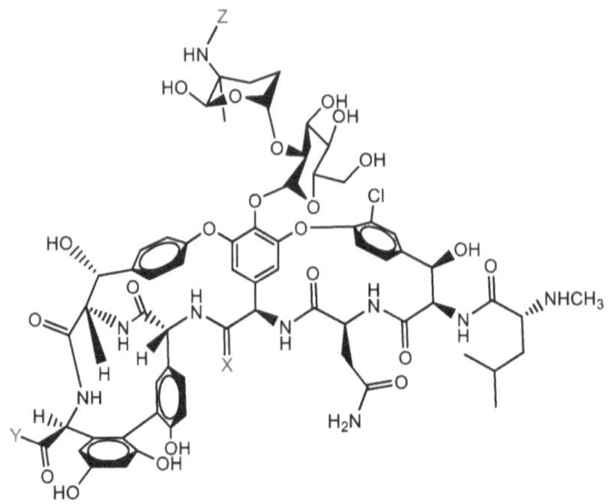

	Vancomycin	#4	#8	#13	#18
X	O	NH	NH	H_2	H_2
Y	OH	OH	OH	$NHC_3H_6N^+(CH_3)_3$	$NHC_3H_6N^+(CH_3)_3$
Z	H	H	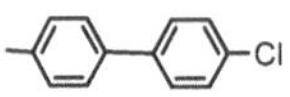	H	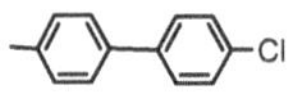
MIC#	250	31	0.13	0.16	0.01

#) MIC is an indicate of the antimicrobial potency in μg/mL, and the smaller number means the higher potency. See text for details.

Fig. 1.7.1 Vancomycin and their modified compounds with enhanced antimicrobial potency.

(μg/mL) shown in Fig. 1.7.1, which is required for VanA vancomycin-resistant *Enterococci* (VRE) (*Enterococcus faecalis* ATCC BAA-2317).

The compounds #4, #8, #13, and #18 showed the antimicrobial potency as high as 8, 1900, 1600, and 25,000 times than that of vancomycin, respectively.

The modification of X-position in Fig. 1.7.1 enhances dual D-Ala-D-Ala/D-Lac binding and inhibits transpeptidase and cell wall biosynthesis. The modification of Y-position in Fig. 1.7.1 induces cell membrane permeability. The modification of Z-position in the figure inhibits transglycosylase and cell wall biosynthesis.

Unfortunately, methylene substitution (X position) is difficult, so that triply modified vancomycin requires 30 steps and is not easy, industrially speaking. However, industrial chemists can overcome the problem.

1.8 HOW DO TARDIGRADES SURVIVE DESICCATION?

Tardigrades, also called water bears or moss piglets (see Fig. 1.8.1), can survive extreme conditions such as desiccation, high temperature, low pressure, etc., especially when they are in a state of cryptobiosis. They can also survive in radiation of >1000 times compared to usual animals. They can withstand 5000 Gy (gray) of gamma rays—for humans, just 5–10 Gy is lethal. The reason for this may be that tardigrades have a repair system for DNA damage.

How they survive desiccation has been a mystery for over 250 years. The disaccharide trehalose has been proposed (Hengherr et al. 2008). However, some desiccation-resistant tardigrades do not need this sugar, nor do they make it. Nowadays, the use of trehalose is unclear. In addition to trehalose, tardigrade-specific intrinsically disordered proteins (TDPs) appeared as the material increasing torrerance to desiccation (Hincha and Thalhammer, 2012).

Boothby et al. (2017) found that TDP genes are highly expressed during desiccation in multiple tardigrade species. TDPs are required for tardigrade

Fig. 1.8.1 A SEM image of a tardigrade (*Milnesium tardigradum*). *(From Wikipedia. Copyright: Schokraie, E., Warnken, U., Hotz-Wagenblatt, A., Grohme, M.A., Hengherr, S., et al., 2012. Comparative proteome analysis of* Milnesium tardigradum *in early embryonic state versus adults in active and anhydrobiotic state. PLoS ONE 7(9): e45682. https://doi.org/10.1371/journal.pone.0045682, CC BY 2.5, https://commons.wikimedia.org/w/index.php?curid=22716809).*

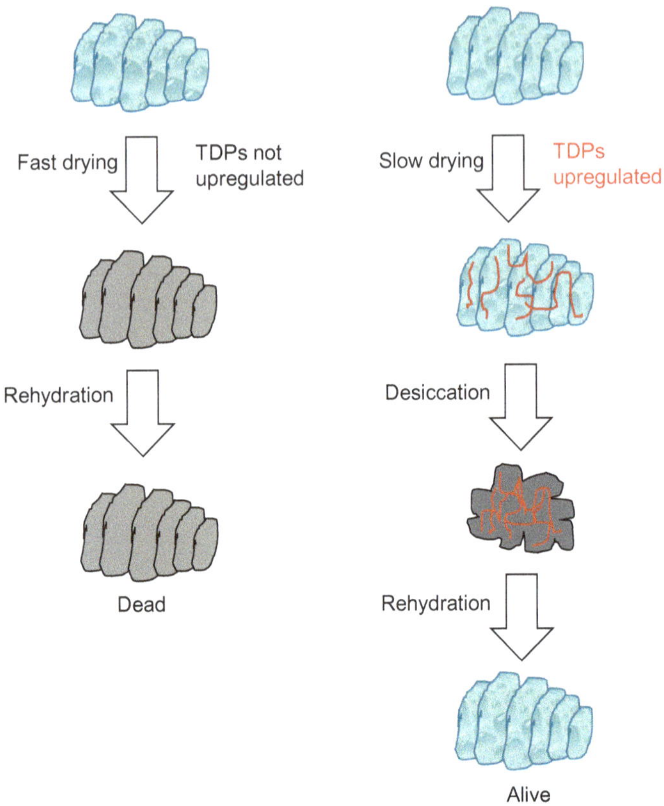

Fig. 1.8.2 Comparison between fast drying and slow drying of tardigrades.

desiccation tolerance, and these genes are sufficient to increase desiccation tolerance when expressed in heterologous systems. During desiccation, TDPs form noncrystalline, amorphous, glassy materials that help their desiccation resistance. When the tardigrades are dried slowly, tardigrade intrinsically disordered proteins (TDPs) are expressed during desiccation (see Fig. 1.8.2). TDPs increase the tolerance against desiccation.

Boothby et al. (2017) concluded that TDPs are functional mediators against desiccation. This explains the requirement of slow drying for tardigrade to survive.

REFERENCES

Ballatore, D.M., Huryn, A., Smith 3rd, B., 2013. Carboxylic acid (bio)isosteres in drug design. Chem. Med. Chem. 8, 385–395.

Bauduin, P., Prevost, S., Farràs, P., Teixidor, F., Diat, O., Zemb, T., 2011. A theta-shaped amphiphilic Cobaltabisdicarbollide anion: Transition from monolayer vesicles to micelles. Angew. Chem. Int. Ed. 50, 5298–5300.

Boothby, T.C., Tapia, H., Brozena, A.H., Piszkiewicz, S., Smith, A.E., Giovannini, I., et al., 2017. Tardigrades use intrinsically disordered proteins to survive desiccation. Mol. Cell 65, 975–984.

Brooks, W.L.A., Sumerlin, B.S., 2016. Synthesis and applications of boronic acid containing polymers: from materials to medicine. Chem. Rev. 116, 1375–1397.

Brown, H.C., 1972. Boranes in Organic Chemistry. Cornell University Press, Ithaca, pp. 346–347.

Brusselle, D., Bauduin, P., Girard, L., Zaulet, A., Viñas, C., Teixidor, F., Ly, I., Diat, O., 2013. Lyotropic lamellar phase formed from Monolayered θ-shaped Carborane-cage Amphiphiles. Angew. Chem. Int. Ed. 52, 12114–12118.

Bunn, C.W., Howells, E.R., 1954. Structures of molecules and crystals of fluoro-carbons. Nature 174, 549–551.

Cabrea-Gonzalez, J., Sanchez-Arderiu, V., Vinas, C., Parella, T., Teixidor, F., Nunez, R., 2016. Redox-active metallacarborane-decorated octasilsesquioxanes. Electrochemical and thermal properties. Inorg. Chem. 55, 11630–11634.

Cordes, D.B., Lickiss, P.D., Rataboul, F., 2010. Recent developments in the chemistry of cubic polyhedral oligosilsesquioxanes. Chem. Rev. 110, 2081–2173.

Dong, X., Oganov, A.R., Goncharov, A.F., Stavrou, E., Lobanov, S., et al., 2017. A stable compound of helium and sodium at high pressure. Nat. Chem. 9, 440–445.

Draganov, A., Wang, D., Wang, B., 2014. The future of boron in medicinal chemistry: therapeutic and diagnostic applications. In: Schwarz, J. (Ed.), Atypical Elements in Drug Design. Springer, pp. 1–27. https://doi.org/10.1007/7355_2014_65.

Farràs, P., Juárez-Pérez, E.J., Lepsik, M., Luque, R., Núñez, R., Teixidor, F., 2012. Metallacarboranes and their interactions: theoretical insights and their applicability. Chem. Soc. Rev. 41, 3445–3463.

Fischer, E.O., Pfab, W., 1952. Zur Kristallstruktur der Di-Cyclopentadienyl-Verbindungen des zweiwertigen Eisens, Kobalts und Nickels. Z. Naturforsch. B 7, 377–379.

Fujimori, T., dos Santos, H., Endo, M., Kanek, K., Tománek, D., 2013. Formation and properties of selenium double-helices inside double-wall carbon nanotubes: experiment and theory. ACS Nano 7, 5607–5613.

Grant, J.T., Carrero, C.A., Goeltl, F., Vengas, J., Mueller, P., et al., 2016. Selective oxidative dehydrogenation of propane to propene using boron nitride catalysts. Science 354, 1570–1573.

Hengherr, S., Heyer, A.G., Köhler, H.-R., Schill, R.O., 2008. Trehalose and anhydrobiosis in tardigrades—evidence for divergence in responses to dehydration. FEBS J. 275, 281–288.

Hincha, D.K., Thalhammer, A., 2012. LEA proteins: IDPs with versatile functions in cellular dehydration tolerance. Biochem. Soc. Trans. 40, 1000–1003.

Ivanov, A.S., Morris, A.J., Bozhenko, K.V., Pickard, C.J., Boldyrev, A.I., 2012. Inorganic double-helix structures of unusually simple lithium-phosphorus species. Angew. Chem. Int. Ed. 51, 8330–8333.

Kealy, T.J., Pauson, P.L., 1951. A new type of Organo-Iron compound. Nature 168, 1039–1040.

Li, C., Wang, J., Barton, L.M., Yu, S., Tian, M., et al., 2017. Decarboxylative borylation. Science. https://doi.org/10.1126/science.aam7355.

Miller, S.A., Tebboth, J.A., Tremaine, J.F., 1952. 114. Dicyclopentadienyliron. J. Chem. Soc. 632–635.

Noda, H., Furutachi, M., Asada, Y., Shibasaki, M., Kumagai, N., 2017. Unique physicochemical and catalytic properties dictated by the B_3NO_2 ring system. Nat. Chem. 9, 571–577.

Okano, A., Isley, N.A., Boger, D.L., 2017. Peripheral modifications of [Ψ[CH_2NH]Tpg[4]] vancomycin with added synergistic mechanisms of action provide durable and potent antibiotics. PNAS 114, E5052–E5061.
Pfister, D., Schäfer, K., Ott, C., Cerke, B., Pötttgen, R., et al., 2016. Inorganic double helices in semiconducting SnIP. Adv. Mater. 28, 9783–9791.
Qin, T., Cornella, J., Li, C., Malins, L.R., Edwards, J.T., et al., 2016. A general alkyl-alkyl cross-coupling enabled by redox-active esters and alkylzinc reagents. Science 352, 801–805.
Sharma, J., Chhabra, R., Cheng, A., Brownell, J., Liu, Y., Yan, H., 2009. Control of self-assembly of DNA tubes through integration of gold nanoparticles. Science 323, 112–116.
Sivaev, I.B., Bregadze, V.I., 1999. Chemistry of cobalt bis (dicarbollides). A review. Collect. Czech. Chem. Commun. 64, 783–805.
Soghomonian, V., Chen, Q., Haushalter, R.C., Zubieta, J., O'Connor, C.J., 1993. An inorganic double helix: hydrothermal synthesis, structure, and magnetism of chiral $[(CH_3)_2NH_2]K_4[V_{10}O_{10}(H_2O)_2(OH)_4(PO_4)_7]\cdot 4H_2O$. Science 1993 (259), 1596–1599.
Srivastava, S., Santos, A., Critchley, K., Kim, K.S., Podsiadlo, P., et al., 2010. Light-controlled self-assembly of semiconductor nanoparticles into twisted ribbons. Science 327, 1355–1359.
Stock, A., Pohland, E., 1926. Borwasserstoffe, VIII. Zur Kenntnis des B_2H_6 und des B_5H_{11}. Berichte (in German) 59, 2210–2215.
Suzuki, A., 2011. Cross-coupling reactions of organoboranes: an easy way to construct CC bonds (Nobel lecture). Angew. Chem. Int. Ed. 50, 6722–6737.
Tang, Z., Zhang, Z., Wang, Y., Glotzer, S.C., Kotov, N.A., 2006. Self-assembly of CdTe nanocrystals into free-floating sheets. Science 314, 274–278.
Toriyama, F., Cornella, J., Wimmer, L., Chen, T.-G., Dixon, D.D., et al., 2016. Redox-active esters in Fe-catalyzed C-C coupling. J. Am. Chem. Soc. 138, 11132–11135.
Trippier, P.C., McGuigan, C., 2010. Boronic acids in medicinal chemistry: anticancer, antibacterial and antiviral applications. MedChemComm 1, 183.
Wang, J., Qin, T., Chen, T.-G., Wimmer, L., Edwards, J.T., et al., 2016. Nickel-catalyzed cross-coupling of redox-active esters with boronic acids. Angew. Chem. Int. Ed. 55, 9676–9679.
Watson, J.D., Crick, F.H., 1953. A structure for deoxyribose nucleic acid. Nature 171, 737–738.
Wilkinson, G., Rosenblum, M., Whiting, M.C., Woodward, R.B., 1952. The structure of iron bis-cyclopentadienyl. J. Am. Chem. Soc. 74, 2125–2126.
Zhao, M.-Q., Zhang, Q., Tian, G.-L., Wei, F., 2014. Emerging double helical nanostructures. Nanoscale 6, 9339–9354.

PART TWO

Fundamental Knowledges and Techniques in Biochemistry

CHAPTER TWO

Fundamental Knowledges and Techniques in Biochemistry

Contents

2.1 AMINO ACIDS

Amino acids, especially 2-, alpha-, or α-amino acids, are biologically important. These prefixes indicate the type of amino acids that have amine ($-NH_2$) and carboxylic acid ($-COOH$) at the same carbon. Therefore, the general chemical formula of the amino acids becomes $H_2NCHRCOOH$, where R is an organic substituent known as a side chain.

Figs. 2.1.1 and 2.1.2 show typical amino acids. Histidine, threonine, isoleucine, tryptophan, leucine, lysine, valine, methionine, and phenylalanine are essential amino acids for humans. These amino acids cannot be synthesized in the human body. The essential amino acids are different from each living thing.

Table 2.1.1 lists abbreviations for 20 amino acids in proteins. Each amino acid has a three-letter abbreviation and a one-letter abbreviation. There are four kinds of side chains: negative, positive, uncharged polar, and nonpolar.

Biochemistry for Materials Science
https://doi.org/10.1016/B978-0-12-817054-0.00002-3

© 2019 Elsevier Inc.
All rights reserved.

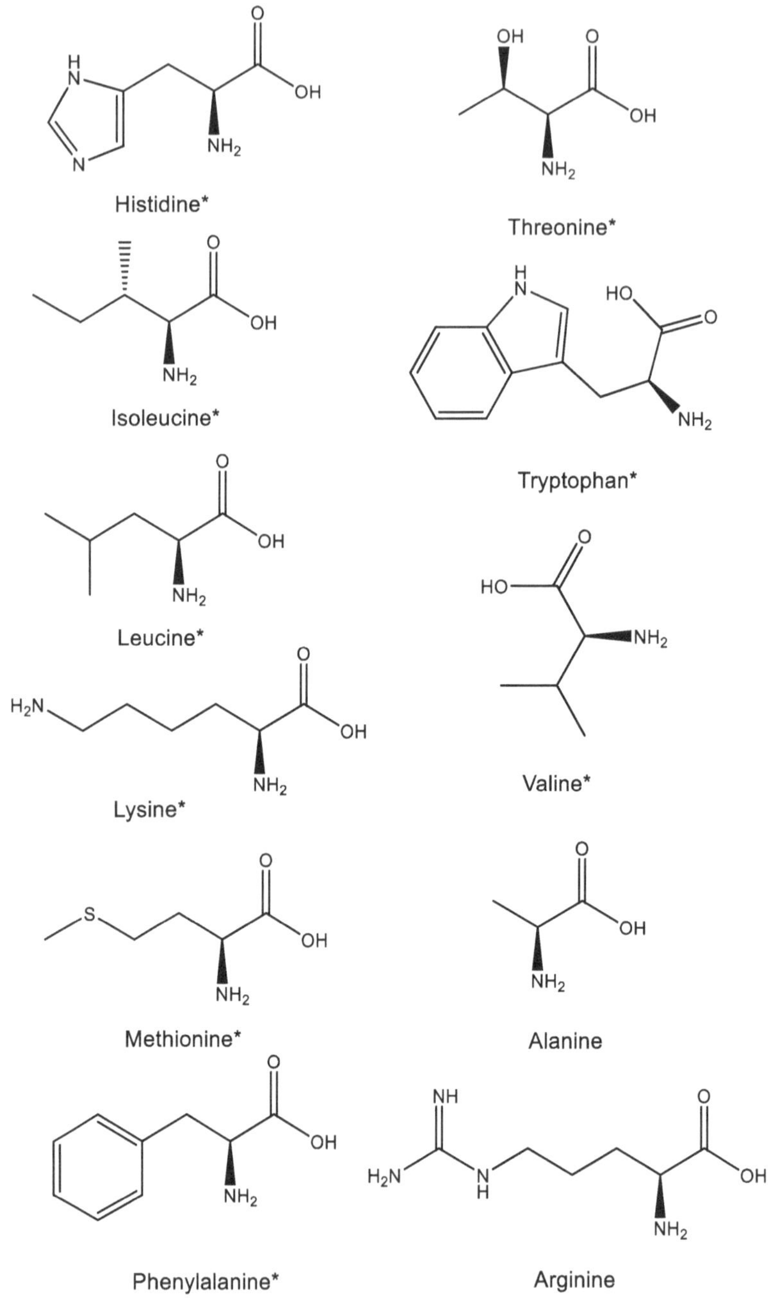

Fig. 2.1.1 Amino acids. (Amino acids with "*" are essential amino acids for human.)

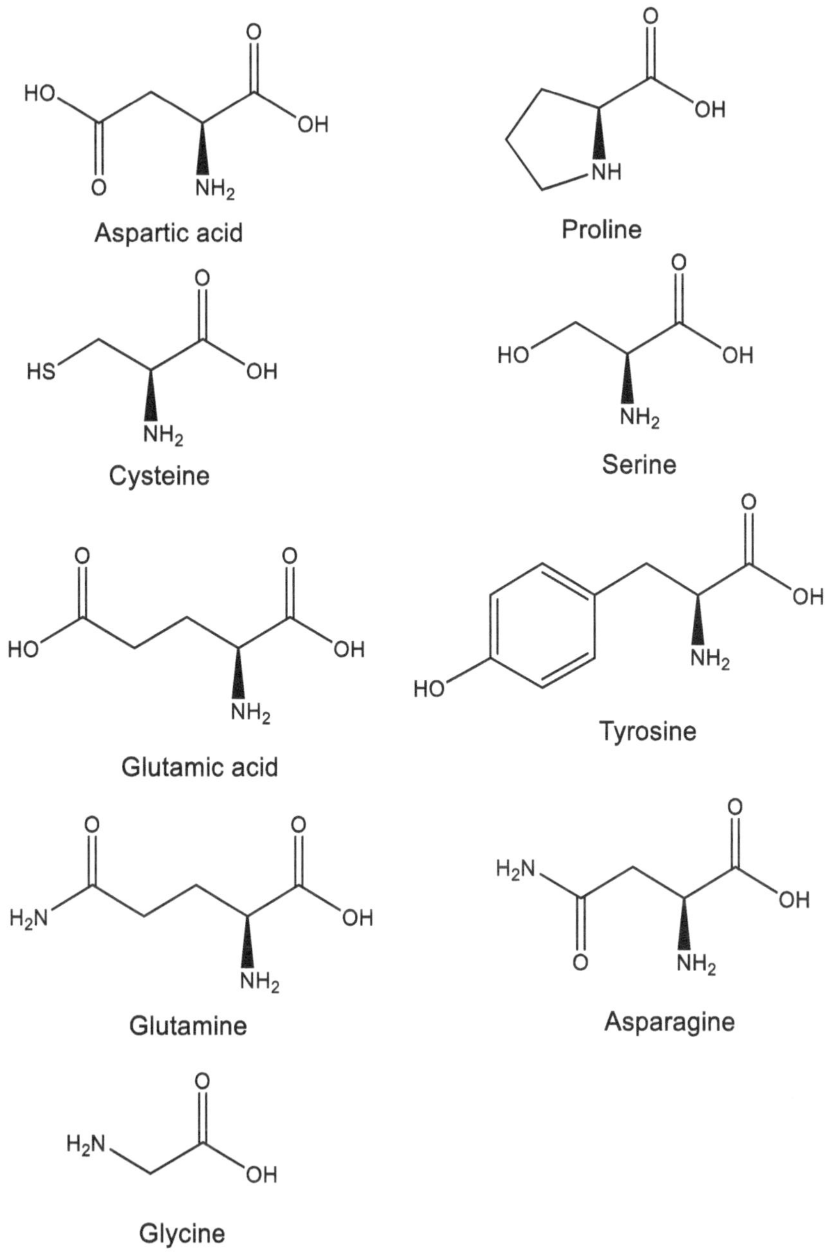

Fig. 2.1.2 Amino acids (part 2).

Table 2.1.1 The 20 Amino Acids (Each Amino Acid Has a Three-Letter and an Alphabet Abbreviation. Side Chains have Negative, Positive, Uncharged Polar, and Nonpolar Characteristics.)

Amino Acid	Three-Letter	Alphabet	Side Chain
Aspartic acid	Asp	D	Negative
Glutamic acid	Glu	E	Negative
Arginine	Arg	R	Positive
Lysine	Lys	K	Positive
Histidine	His	H	Positive
Asparagine	Asn	N	Uncharged polar
Glutamine	Gln	Q	Uncharged polar
Serine	Ser	S	Uncharged polar
Threonine	Thr	T	Uncharged polar
Tyrosine	Tyr	Y	Uncharged polar
Alanine	Ala	A	Nonpolar
Glycine	Gly	G	Nonpolar
Valine	Val	V	Nonpolar
Leucine	Leu	L	Nonpolar
Isoleucine	Ile	I	Nonpolar
Proline	Pro	P	Nonpolar
Phenylalanine	Phe	F	Nonpolar
Methionine	Met	M	Nonpolar
Tryptophan	Trp	W	Nonpolar
Cysteine	Cys	C	Nonpolar

2.2 PROTEINS

Two amino acids with side chains of R_1 and R_2 can combine as follows:

$$\begin{aligned}&NH_2-CH(R_1)-COOH+NH_2-CH(R_2)-COOH\\&\rightarrow NH_2-CH(R_1)-CO-NH-CH(R_2)-COOH+H_2O\end{aligned} \tag{2.2.1}$$

This —CO—NH— bond is called a peptide bond. Amino acids can combine infinitely (see Fig. 2.2.1 for a combination of three amino acids in 3D).

Proline is not an amino acid, but an imino acid. However, it is treated as the amino acid. As proline has an unusual ring structure, in which the peptide bond becomes not as usual:

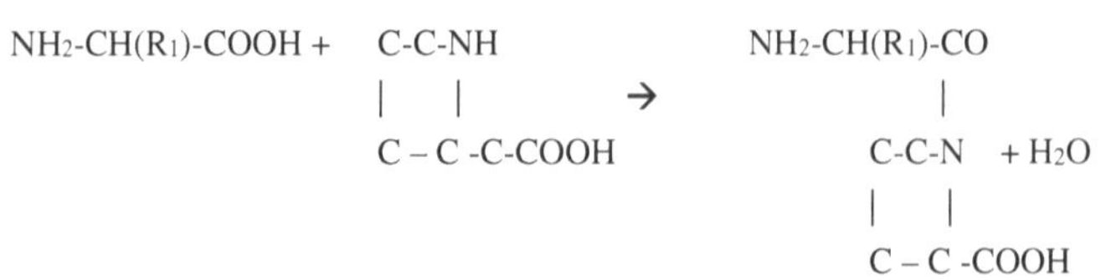

(2.2)

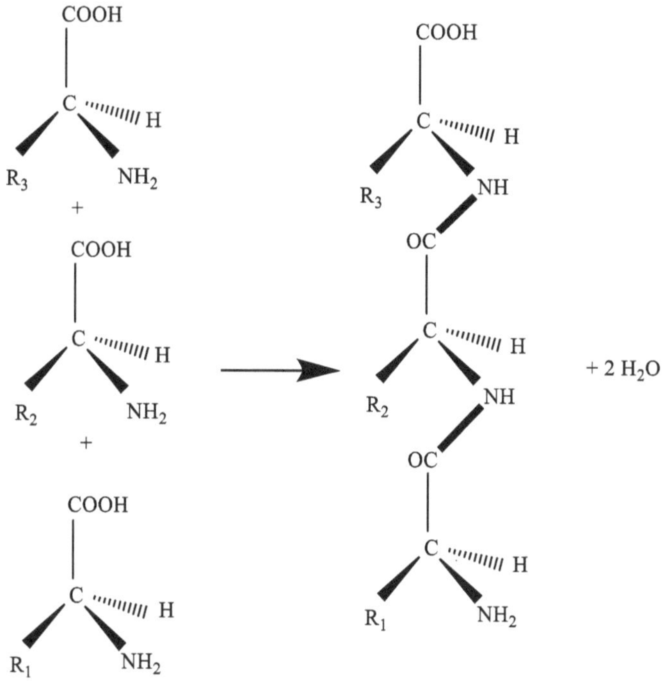

Fig. 2.2.1 Combination of three amino acids by peptide bonds.

The polypeptide polymer molecule is called protein, and some polypeptides have special biological functions. The end of the protein with a free carboxyl group is called the C-terminus or carboxyl terminus (the amino acid with R_3 in Fig. 2.2.1), and the end with a free amino group is called the N-terminus or amino terminus (or the amino acid with R_1 in Fig. 2.2.1).

There are two secondary structures in proteins. One is coil structures, called alpha-helix (α-helix). The other is a motif in proteins, called beta-sheet (β-sheet) structure. The β-sheet is shown as a widearrow.

2.3 THE PROTEIN DATA BANK (PDB)

The name of proteins is registered in RCSB Protein Data Bank (PDB) with PDB ID. When the PDB ID is known, a three-dimensional image of the protein is also registered with the reference(s), which can be obtained by accessing http://www.rcsb.org/pdb/home/home.do.

For example, sperm whale myoglobin by Watson (1969) is registered as "1MBN." As many as 274 sperm whale myoglobins are registered in PDB in March 2017. Fig. 2.3.1A–E shows the 3D structure of 1MBN. This protein is a metalloprotein, which is made of protein itself and heme (or heam). The heme is shown as connected balls in Fig. 2.3.1.

Fig. 2.3.1A–E are enlarged images where the heme is enveloped, in which the polypeptides are shown as ribbon (A–D) or wire (E) licorice. The amino acids reacted with the heme are called active sites and numbered from the N-terminus. For example, F43, H64, V68, and H93 are 43rd, 64th, 68th, and 93rd of phenylalanine, histidine, valine, and histidine, respectively. This information is all available from RCSB PDB.

As shown in Fig. 2.3.1A–C, the structure of myoglobin is like a "hot dog" with a bread roll and a sausage. The bread corresponds to the protein,

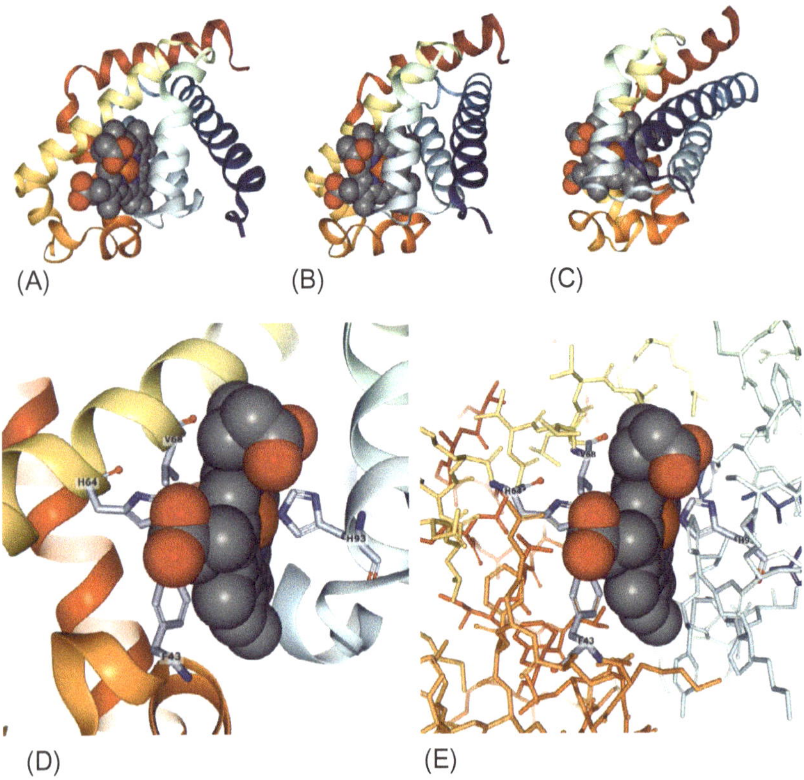

Fig. 2.3.1 Stereoscopy of sperm whale myoglobin (1MBN; Watson (1969)). Gray balls indicate the heme, made of oxygen, an iron, nitrogen, and carbons. Stereograms are available with the combination of (A) and (B), or (B) and (C). The figures of (D) and (E) are enlarged images where the haem is enveloped.

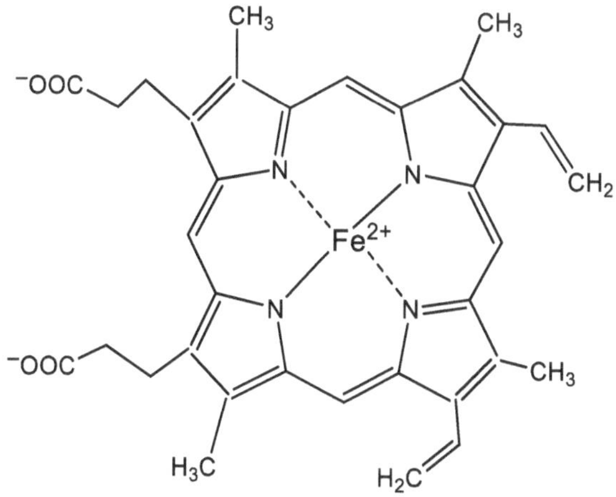

Fig. 2.3.2 The chemical formula of heme (porphyrin IX containing Fe^{2+} ion).

which holds the sausage in a manner comparable to heme (see Fig. 2.3.2). Coils in structures of Fig. 2.3.1A–C, which are colored in the rainbow, are alpha-helix (α-helix).

2.4 STEREOSCOPY

The stereoscopy or stereogram (3D-images) can be available with combinations of Fig. 2.3.1A and B, or Fig. 2.3.1B and C. Red, orange, blue, and gray balls indicate oxygen, iron, nitrogen, and carbon, respectively. The balls in Fig. 2.3.1 indicate a protoporphyrin IX containing Fe, which is called heme (or heam). The heme is made of a porphyrin ring with an iron in the center (see Fig. 2.3.2).

In order to obtain the stereoscopy of the images of Fig. 2.3.1A and B, look at Fig. 2.3.1A with the left eye, and Fig. 2.3.1B with the right eye. Then try to fuse both figures into one. When the two figures merge, the fused figure should become a 3D figure. For those who wear glasses, the stereopsis occurs more easily if you remove your glasses. Seeing the same point of each figure (e.g., the orange ball) with each eye also makes it easier to see the fused image.

2.5 PURINES AND PYRIMIDINES

Fig. 2.5.1A and B shows purines and pyrimidines. Adenine (A) and guanine (G) are purines, and cytosine (C), thymine (T), and uracil

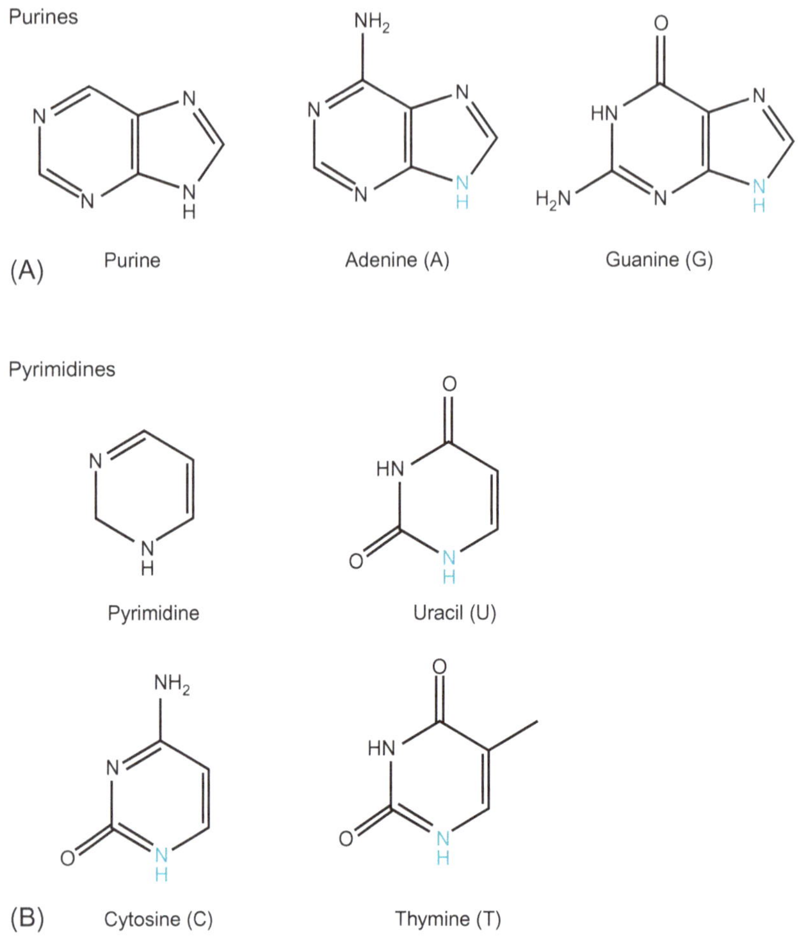

Fig. 2.5.1 Purines and pyrimidines: (A) purines; (B) pyrimidines.

(U) are pyrimidines. These are the most important parts in the nucleic acid, and genetic information is stored in the sequence of these molecules. The gray-colored nitrogen atom (see Fig. 2.5.1A and B) is connected to pentose sugar.

In biochemistry, the abbreviations A, G, C, T, and U are often used to show the genetic information in deoxyribonucleic acid (DNA, see Fig. 2.5.2A) and ribonucleic acid (RNA, see Fig. 2.5.2B). In DNA and RNA, A, G, C, and T, and A, G, C, and U are used, respectively.

These adenine (A) or T, G, and C are bound with the five-carbon sugar, called deoxyribose, which is connected with one phosphoric acid. This single set of molecules is called a nucleotide. When the five-carbon sugar is deoxyribose (Fig. 2.5.2A), the polymer of this molecule is called

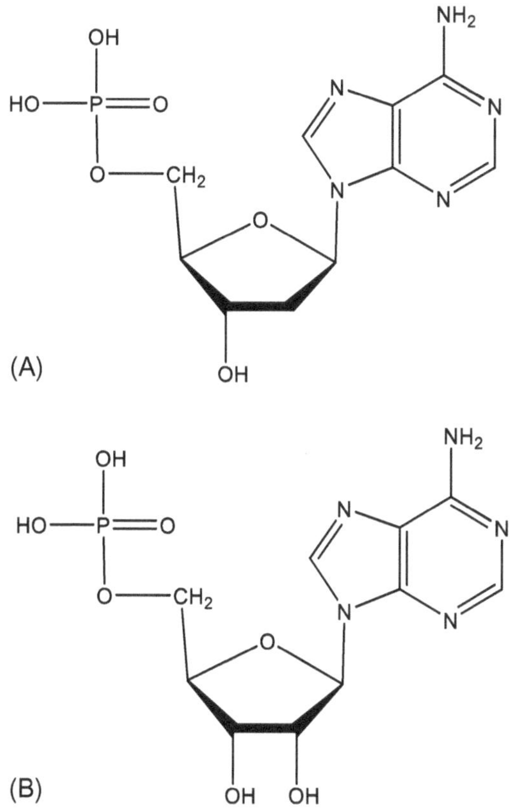

Fig. 2.5.2 Nucleotides: (A) adenine-deoxyribonucleic acid-phosphoric acid; (B) adenine-ribonucleic acid-phosphoric acid. The molecule (B) is also called as adenosine monophosphate (AMP).

deoxyribonucleic acid (DNA). When the five-carbon sugar is ribose (Fig. 2.5.2B), the polymer of this molecule is called ribonucleic acid (RNA).

2.6 HYDROGEN BOND

The five molecules form a constant combination by the hydrogen bond. Adenine (A) combines with thymine (T), or uracil (U) in RNA, and guanine (G) combines with cytosine (C). The hydrogen bonds are shown as dotted lines in Fig. 2.6.1. The hydrogen bond is electrostatic attraction between polar molecules, such as nitrogen or oxygen and hydrogen. The hydrogen bonds appear in various reactions in living things. The 3D structure of proteins, which is a hot topic in biochemical research, is

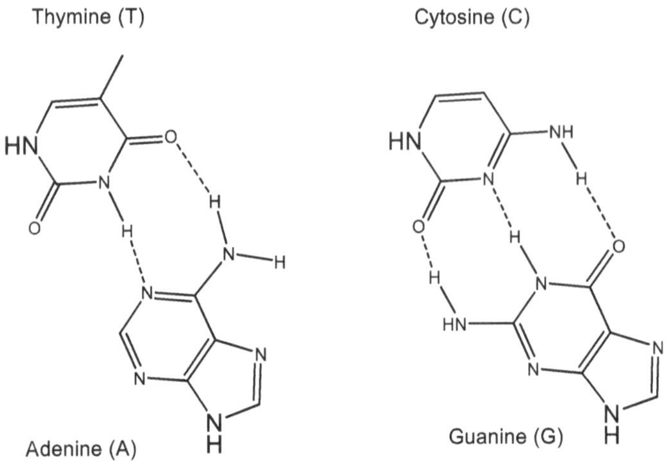

Fig. 2.6.1 Hydrogen bonds.

defined using hydrogen bonds. (For this area of research, supercomputers are needed to simulate the 3D structure of proteins.)

It is astonishing that living things use the hydrogen bonds to store the genetic information only by A-T and G-C pairs. These bonds are not strong enough to resist being cut, but they are strong enough not to be cut easily. Some carcinogens have similar structure to these molecules, and enter into the A-T or G-C pairs and break the genetic information.

2.7 HELIX AND DOUBLE HELIX

Nucleotides make long polymers, as shown in the left side of Fig. 2.7.1, from 5′ to 3′. The two polymers makes the double helix, as shown in Fig. 2.7.1. Two ends of phosphoric acid and deoxyribose are called the 5′ end and 3′ end, respectively. The direction of 5′→3′ is defined as the normal direction, and when the base direction of the DNA in Fig. 2.7.1 needs to be shown, it is shown along the normal direction as AGT, not as TGA.

2.8 SYNTHESIS OF PROTEINS FROM DNA

There are two steps for protein synthesis from DNA: DNA to RNA (step 1); and RNA to protein (step 2). This section explains these two steps.

For the synthesis of proteins from DNA, the proteins are not directly synthesized from DNA, but copied through RNA (transcription); the

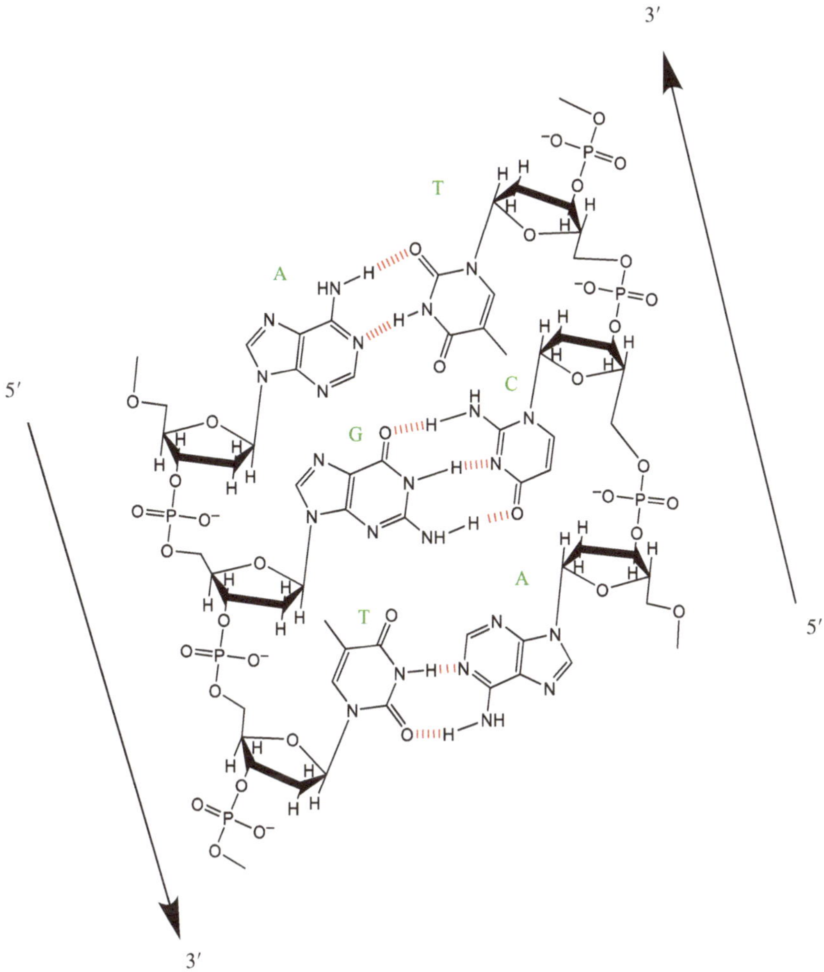

Fig. 2.7.1 DNA characters of A, G, T, and C indicate adenine, guanine, thymine, and cytocine. Dotted bonds show hydrogen bonds.

protein is then synthesized from the RNA (translation) (see Fig. 2.8.1). The genetic information of DNA (Fig. 2.8.1A) is copied to RNA (Fig. 2.8.1B), which is called "transcription." It should be noted that RNA is a chain of ribonucleotides, T (thymine) in DNA is transcribed into U (uracil) in RNA, and RNA is not a double helix molecule, but a single line molecule. Compared to DNA molecules that are 250 million nucleotide-pairs long, RNA molecules are no more than a few thousand nucleotides long. Many RNAs are much shorter.

The information of RNA is then copied to protein, which is called "translation." Three sets of nucleic acids are called a codon, which is

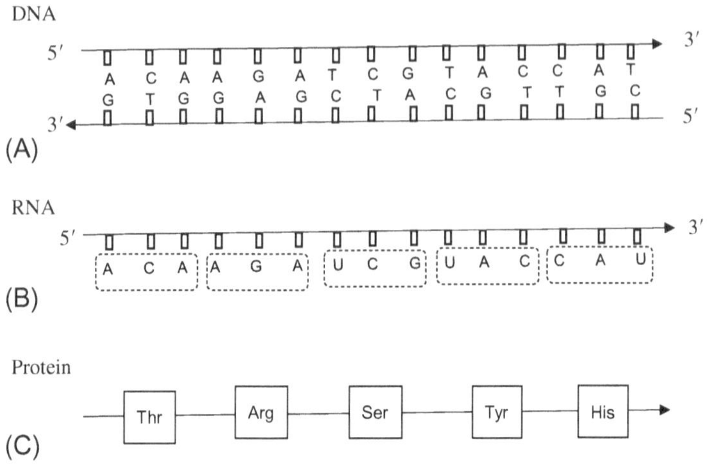

Fig. 2.8.1 Schematic diagrams of DNA(A) to RNA(B), and RNA(B) to protein(C).

indicated as dotted squares in Fig. 2.8.1B. Each codon is translated into one protein following the genetic code (see Fig. 2.8.1C and Table 2.8.1).

In the following sections, the two processes of transcription and translation are explained in more detail.

Table 2.8.1 The Genetic Code

Asp	GAU, GAU
Glu	GAA, GAG
Arg	AGA,AGG,CGA,CGC,CGG,CGU
Lys	AAA, AAG
His	CAC, CAU
Asn	AAC, AAU
Gln	CAA, CAG
Ser	AGC, AGU, UCA, UCC, UCG, UCU
Thr	ACA, ACC, ACG, ACU
Tyr	UAC, UAU
Ala	GCA, GCC, GCG, GCU
Gly	CAA, CAG
Val	GUA, GUC, GUG, GUU
Leu	UUA, UUG, CUA, CUC, CUG, CUU
Ile	AUA, AUC, AUU
Pro	CCA, CCC, CCG, CCU
Phe	UUC, UUU
Met	AUG
Trp	UGG
Cys	UGC, UGU
stop	UAA, UAG, UGA

2.9 TRANSCRIPTION (SYNTHESIS OF RNA FROM DNA)

Transcription is performed by an enzyme named RNA polymerase. The RNA polymerase moves stepwise along the DNA, unwinding the DNA helix in an mRNA chain 5′- to 3′- direction. As the RNA is released from the DNA in a short time, the RNA syntheses can start at various positions of the DNA before the RNA syntheses finish.

The genetic information in RNA is not stored permanently, so that one mistake occurs in every 10^4 nucleotide copies. In contrast, in DNA, one mistake occurs in every 10^7 nucleotide copies. There are several types of RNAs, which are shown in Table 2.9.1. Most RNAs are rRNAs, and mRNAs account for 3%–5%. miRNAs and siRNAs are key in gene expression in eukaryotes.

Bacteria contain a single type of RNA polymerase, which performs the transcription ofDNA into mRNA. An mRNA molecule is produced when this enzyme initiates transcription at a promotor, synthesizes the RNA by chain elongation, stops transcription at a terminator, and releases both the DNA template and the completed mRNA molecule.

In eukaryotic cells, the process of transcription is more complex, because there are three types of RNA polymerases—polymerase I, II, and III—that are related evolutionarily to one another and to the bacterial polymerase. RNA polymerase II synthesizes eukaryotic mRNA. This enzyme requires

Table 2.9.1 Several Types of RNA

Type of RNA	Function
mRNAs	Messenger RNAs, code for proteins
rRNAs	Ribosomal RNAs, form the basic structure of the ribosome
tRNAs	Transfer RNAs, adaptors between mRNA and amino acids
snRNAs	Small nuclear RNAs, used to modify rRNAs
snoRNAs	Small nucleolar RNAs, used to process and chemically modify rRNAs
scaRNAs	Small cajal RNAs, used to modify snoRNAs and snRNAs
miRNAs	MicroRNAs, regulate gene expression typically by blocking translation of selective mRNAs
siRNAs	Small interfering RNAs, turn off gene expression by directing degradation of selective mRNAs and the establishment of compact chromatin structures
Other noncoding RNAs	Function in diverse cell processes, including telomere synthesis, X-chromosome inactivation, and the transport of proteins into the ER

a series of additional proteins, the general transcription factors, to initiate transcription on a purified DNA template, and still more proteins (including chromatin (complex with protein and DNA)-remodeling complexes and histone (the protein forming chromatin complex)-modifying enzymes) to initiate transcription on its chromatin templates inside the cell.

During the elongation phase of transcription, the nascent RNA undergoes three types of processing events: a special nucleotide is added to its 5′ end (capping); intron sequences (the removed RNA by splicing) are removed from the middle of the RNA molecule (splicing); and the 3′ end of the RNA is generated (cleavage and polyadenylation). Each process is initiated by proteins that travel along with RNA polymerase II by binding to sites on its long, extended C-terminal tail. Splicing is unusual in that many of its key steps are carried out by specialized RNA molecules rather than proteins. Properly processed mRNAs pass through nuclear pore complexes into the cytosol, where they are translated into protein.

For some genes, RNA is the final product. In eukaryotes, these genes are usually transcribed by either RNA polymerase I or RNA polymerase III. RNA polymerase I makes the ribosomal RNAs. After their synthesis as a large precursor, the rRNAs are chemically modified, cleaved, and assembled into the two ribosomal subunits in the nucleolus—a distinct subnuclear structure that also helps to process some smaller RNA-protein complexes in the cell. Additional subnuclear structures (including Cajal bodies—tiny circular nucleolar accessory bodies—and interchromatin granule clusters) are sites where components involved in RNA processing are assembled, stored, and recycled (Alberts et al., 2008).

2.10 SYNTHESIS OF PROTEIN FROM RNA

The translation of the nucleotide sequence of an mRNA molecule into protein takes place in the cytoplasm on a large ribonucleoprotein assembly called a ribosome. The amino acids used for protein synthesis are first attached to a family of tRNA molecules, each of which recognizes, by complementary base-pair interactions, particular sets of three nucleotides in the mRNA (codons). The sequence of nucleotides in the mRNA is then read from one end to the other insets of three according to the genetic code (see Table 2.8.1).

To initiate translation, a small ribosomal subunit binds to the mRNA molecule at a start codon (AUG; see Table 2.8.1) that is recognized by a unique initiator tRNA molecule. A large ribosomal subunit binds to complete the ribosome and begin protein synthesis. During this phase, aminoacyl-tRNAs—each bearing a specific amino acid—bind sequentially to the appropriate codons in mRNA through complementary base pairing between tRNA anticodons and mRNA codons. Each amino acid is added to the C-terminal end of the growing polypeptide in four sequential steps: aminoacyl-tRNA binding, followed by peptide bond formation, followed by two ribosome translocation steps. Elongation factors use GTP hydrolysis to drive these reactions forward and to improve the accuracy of amino acid selection. The mRNA molecule progresses codon by codon through the ribosome in the 5′- to 3′- direction until it reaches one of three stop codons (see Table 2.8.1). A release factor then binds to the ribosome, terminating translation and releasing the completed polypeptide.

Eukaryotic and bacterial ribosomes are closely related, despite differences in the number and size of their rRNA and protein components. The rRNA has the dominant role in translation, determining the overall structure of the ribosome, forming the binding sites for the tRNAs, matching the tRNAs to codons in the mRNA, and creating the active site of the peptidyl transferase enzyme that links amino acids together during translation.

In the final steps of protein synthesis, two distinct types of molecular chaperones guide the folding of polypeptide chains. These chaperones recognize exposed hydrophobic patches on proteins and serve to prevent the protein aggregation that would compete with the folding of newly synthesized proteins into their correct-three-dimensional conformations. This protein folding process must also compete with an elaborate quality control mechanism that destroys proteins with abnormally exposed hydrophobic patches. In this case, ubiquitin is covalently added to a misfolded protein by a ubiquitin ligase, and the resulting polyubiquitin chain is recognized by the cap on a proteasome that moves the entire protein to the interior of the proteasome for proteolytic degradation. A closely related proteolytic mechanism, based on special degradation signals recognized by ubiquitin ligases, is used to determine the lifetimes of many normally folded proteins. By this method, selected normal proteins are removed from the cell (Alberts et al., 2008).

2.11 SYNTHESIS OF PROTEINS WITH HEXAHISTIDINE-TAG (6xHis-TAG)

In biochemistry, proteins are often synthesized with the polyhistidine tag using *Escherichia coli* (*E. coli*), by a popular method for biochemists.

For the synthesis of the target protein, the amino acid sequence of the target protein must be identified. When this sequence is known, the sequence of DNA is also determined, because three DNA sequences determine one amino acid. On this sequence, the six-repetition of histidine (histidine is an essential amino acid, shown in Fig. 2.1.1), called the hexahistidine-tag (or 6xHis-tag or His6 tag), is added.

This DNA is inserted into that of *E. coli*. The DNA of *E. coli*. is one circular DNA, so that the DNA is cut at one place (the place commonly or preferentially used is called a "cloning vector"), and connected again at two places as the 6xHis-tag-(DNA of the target protein)-6xHis-tag, which is called the "recombinant DNA." When the protein is produced (or "expressed") by *E. coli*, the protein (recombinant protein) has the 6xHis-tag at the beginning and the end. This 6xHis-tag is used for purification of protein, described in the next section.

2.12 PURIFICATION OF PROTEIN BY IMAC USING NI-NTA RESIN

The 6xHis-tag has high affinity with Ni^{2+}, because the poly-histidine easily makes the stable chelate-complex with Ni^{2+}. This character is used for purification of the protein. Nickel ions are immobilized by chelation with nitrilotriacetic acid (Ni-NTA) bound to a solid support of chromatographic resin (Ni-NTA resin; see Fig. 2.12.1). This resin is commercially available, and the purification method of proteins with the 6xHis-tag is called immobilized metal affinity chromatography (IMAC).

After the *E. coli* produced the protein based on the recombinant DNA, the protein needs to be purified. The cell membranes of *E. coli* containing the target recombinant protein are broken. Then the solution containing the recombinant protein is passed through the column packed with the Ni-NTA resin at relatively high pH (e.g., ~8) with ~20 mM imidazole (see Fig. 17.5.2). The tagged protein passes through the column, and is adsorbed on the resin, while other materials, such as recombinant DNA,

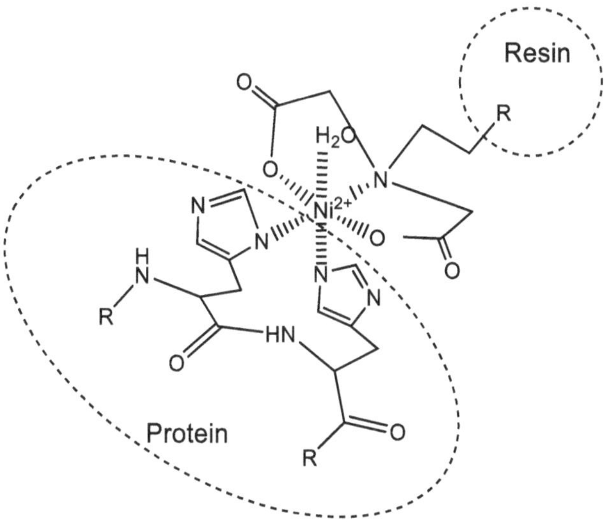

Fig. 2.12.1 Ni-NTA resin with adsorption of the poly-histidine-tagged protein.

cytoplasm, and other proteins and enzymes that are not tagged, are washed away. Then the column is washed with ~200 mM imidazole for the target protein to recover. Finally, the poli-histidine tag is removed by endopeptidases, and pure proteins are recovered.

REFERENCES

Alberts, B., Johnson, A., Lewis, J., et al., 2008. Molecular Biology of the Cell, 6th ed. Garland Science, New York.

Watson, H.C., 1969. The stereochemistry of the protein myoglobin. Prog. Stereochem. 4, 299.

PART THREE

Biochemistry Related Materials and Topics

CHAPTER THREE

New Abiological Metalloenzymes —[M]-porphyrin IX-proteins —

Contents

3.1 INTRODUCTION

This section focuses on metalloenzymes that are abiologic. This means that not transition metals but unusual metals such as platinum group elements are used in metalloenzymes. It is hoped that readers who have only inorganic and/or organic chemical knowledge but are lacking in biochemical knowledge will be able to understand this section. The goal is for the reader to read and understand biochemical reports in relevant journals without difficulty.

The transition metals control the reactivity and selectivity of the substrate in the reaction system. In order to increase reactivity and selectivity, protein engineering has expanded to modify the native metalloenzymes into abiotical metalloenzymes.

One effective method for creating artificial metalloproteins with different reactivity and selectivity involves replacing the metal sites of the original protein with precious-metal complexes. The interactions of the substrate

Biochemistry for Materials Science
https://doi.org/10.1016/B978-0-12-817054-0.00003-5
© 2019 Elsevier Inc.
All rights reserved.

with the protein become apparently different from those of the original protein, because the metal complex determines the substrate binding site.

3.2 THREE-DIMENSIONAL IMAGE OF MYOGLOBIN

Key et al. (2016) used sperm whale myoglobin (Physeter Fcatodon) named 1MBN (Watson, 1969), which is RCSB Protein Data Bank (PDB) ID (see Section 2.3). Three-dimensional images of myoglobin 1MBN can be obtained from the PDB database, and are shown in Fig. 2.3.1.

As shown in Figs. 2.3.1A–C, the structure of myoglobin is like a "hot dog" with a bread roll and a sausage. The bread corresponds to the protein, which pinches the sausage comparable to a heme (Fig. 2.3.2). Coils in structures of Figs. 2.3.1A–C are alpha-helix (α-helix).

Fig. 2.3.1D and E are enlarged images where the heme is enveloped, in which the polypeptides are shown as ribbon or wire (licorice), respectively. The amino acids that have reacted with the heme are called active sites and numbered from the N-terminus. For example, F43, H64, V68, and H93 are 43rd, 64th, 68th, and 93rd of phenylalanine, histidine, valine, and histidine (see Table 2.1.1), respectively.

3.3 SYNTHESIS OF APO-PIX MYOGLOBIN AND METAL-PORPHYRIN IX (M-PIX)

In order to obtain myoglobin with different metals from Fe, such as Rh, Ir, Ru, Mn, or Co-porphyrin IXs, Fe-PIX was synthesized at the beginning, then Fe was replaced into other metals. However, during the changing of the metals, the protein could be damaged, resulting in improper or irreproducible performances as enzymes. Furthermore, the exchange of metals from iron could not be perfect, which can affect the experimental results.

To overcome such irregularities, Key et al. (2016) devised a new synthesis method for the metalloenzymes with variation of metals and proteins as well. Briefly speaking, they synthesized apo-PIX myoglobin ("apo" means "without"—i.e., myoglobin without porphyrin IX and metals) by the biochemical method using *Escherichia coli* (abbreviated as *E. coli*). Separately, they synthesized eight kinds of [M]-PIX called metal-cofactor or [M]-cofactor, where [M] represents transition or precious metals such as Fe, Co, Cu, Mn, Rh, Ir, Ir(Me), Ru(CO), or Ag. Finally, they mixed apo-PIX myoglobin and [M]-PIX in a stoichiometric fashion. Fig. 3.3.1 shows a schematic diagram of the experiment.

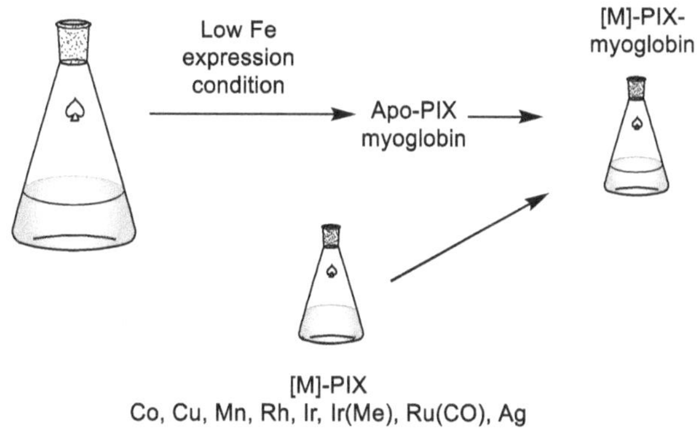

Fig. 3.3.1 A schematic diagram of the experiment of Key et al. (2016). At the beginning, *Escherichia coli* produces the apo-PIX myoglobin, which does not contain any apo-PIX (metal-cofactors as well as any metals). The apo-PIX myoglobin is then purified. [M]-PIX, of which [M] is either the transition metal or precious metal of Fe, Co, Cu, Mn, Rh, Ir, Ir(Me), Ru(CO), or Ag, is synthesized separately. The apo-PIX myoglobin and the [M]-PIX are then mixed in stoichiometric fashion. Thus each [M]-PIX myoglobin is synthesized. When the apo-PIX myoglobin with the modified amino acid is used, the different protein is expressed by *E. coli.*

Table 3.3.1 Turnover numbers (see text) in the reactions of (A) and (B), which are shown in Fig. 3.4.1A and B, respectively

	93H	93C	93D	93E	93M	93S	93A	93G
(A) C—H insertion reaction								
Fe(Cl)-PIX								
Co(Cl)-PIX	+							
Cu-PIX								
Mn(Cl)-PIX								
Rh-PIX	+					+		
Ir(Cl)-PIX	+					+		
Ir(Me)-PIX	+++	+++	+++	+++	++	+++	+++	+++
Ru(CO)-PIX						+		
Ag-PIX								
(B) Cyclopropanation reaction								
Fe(Cl)-PIX								
Co(Cl)-PIX								
Cu-PIX								
Mn(Cl)-PIX								
Rh-PIX								
Ir(Cl)-PIX	+	+		+	+	+	+	+
Ir(Me)-PIX	++	+++		+++	+++	+++	+++	+++
Ru(CO)-PIX								
Ag-PIX								

The horizontal and vertical arrays indicate the apo-PIX myoglobin with the modified amino acid at the 93rd site and [M]-PIX cofactor, respectively. The signs of "+++", "++" and "+" mean turnover number (of >60, 60–31, and 30–11, and >20, 20–11, and 10–6 for (A) and (B), respectively.

In addition, Key et al. modified the key amino acid in myoglobin at the 93rd site from H (histidine) to C, D, E, M, S, A, and G (see Table 2.1.1 for the alphabetical abbreviation of the amino acid), into eight kinds in total. Thus they could synthesize eight apo-PIX myoglobins and nine [M]-cofactors. As a result, they could test 72 (=8 × 9) combinations of [M] and the key amino acids without damaging the myoglobin. Table 3.3.1 shows combinations of the key amino acids.

3.4 EVALUATION OF THE PERFORMANCE OF METALLOPROTEINS AS CATALYSTS

The performance of the array of artificial [M]-myoglobins containing various metals and axial ligands (Table 3.3.1) was evaluated as catalysts for the reaction of diazoester (**1**) to form chiral dihydrobenzofuran (**2**) (Fig. 3.4.1A). The "turnover number" (TON) shown in Table 3.3.1 is the number of moles of substrate of which a mole of catalyst can convert before becoming inactivated. The larger the TON is, the better the catalyst is. The ideal catalyst would have an infinite TON.

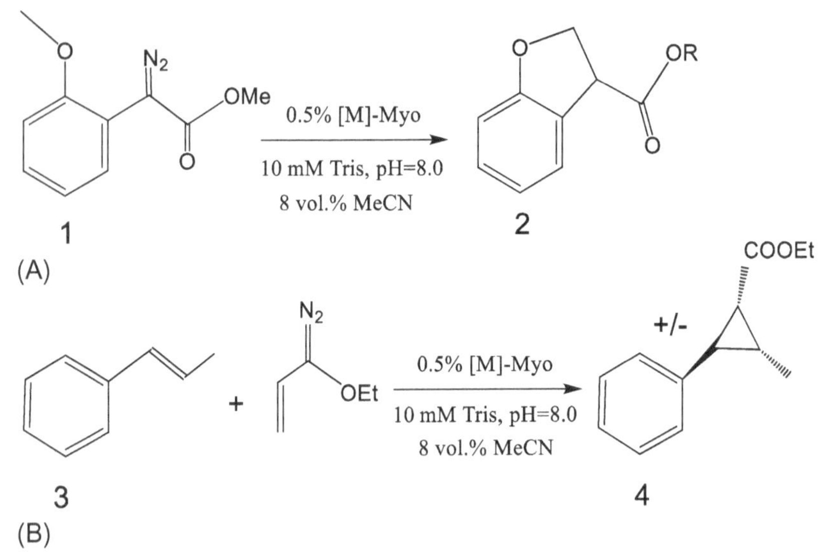

Fig. 3.4.1 Reactions for evaluation of performances of [M]-myoglobins. (A) C—H insertion reaction, which is a reaction of diazoester (**1**) forming into chiral dihydrobenzofuran (**2**). (B) Cyclopropanation reaction, which is addition of carbenes (**4**) to an internal olefin (**3**).

The most active catalysts were those containing Ir(Me)-PIX, which were formed by incorporating both an abiological metal (Ir) and an abiological axial ligand (-CH_3) that cannot be incorporated by the standard mutagenesis techniques. ("Mutagenesis" is when genetic information is changed, resulting in mutation. Standard mutagenesis techniques include site-directed mutagenesis but not ligand–metal exchange into an abiological metal such as Ir(Me).) The eight myoglobins containing Ir(Me)-PIX formed enantio-enriched dibenzohydrofuran (**2**) in up to 50% TON before any further mutagenesis.

In summary, the method for producing the metalloenzyme developed in Sections 3.1–3.4 is as follows: (i) selecting the target reaction and initial metalloenzyme; (ii) choosing the metal with the highest performance; and (iii) changing the amino acids at the reaction sites with the best performance.

3.5 INTRODUCTION OF CYTOCHROME P450 (CYP119) METALLOENZYME

In this section, the metalloenzyme synthesized by Hartwig's group in Berkeley, University of California is in the spotlight again. In this case, they used a cytochrome P450 protein for the starting metalloenzyme (Dydio et al., 2016). The P450 enzymes constitute a superfamily of hemebinding monooxygenases involved in many kinds of biosynthetic pathways, catalyzing reactions that include chemo-, regio-, stereo-, and site-selective C—H hydroxylations of natural complex products (Sigel et al., 2007).

The structure of P450 (CYP119) from Protein Data Bank 1IO7 (Park et al., 2002) is shown in Fig. 3.5.1. The heme, which contains an Fe atom, is shown as the center balls of Fig. 3.5.1. Fig. 3.5.2 shows the enlarged figure of Fig. 3.5.1 around the heme. Only the amino acids reacted with the heme are plotted in Fig. 3.5.2. The number from the N-terminus and the species of amino acids are depicted in the figure.

Dydio et al. (2016) chose CYP119 from P450 families. The CYP119 protein is a metalloenzyme from *Sulfolobus solfataricus*, which lives in volcanic springs with high temperature (75–80°C) and high pH (2–3) conditions. CYP119 has higher melting temperatures (69°C) than those of P450-BM3 (45°C) or P450-CAM (40°C). Therefore, this protein is suited to catalytic reactions.

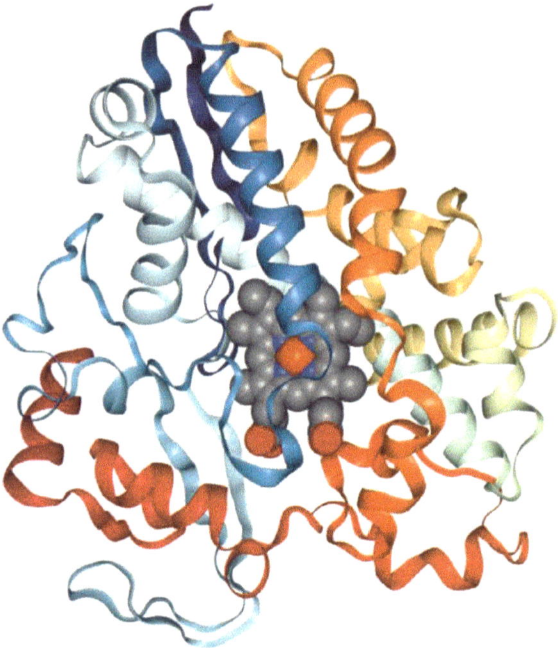

Fig. 3.5.1 The structure of P450 (CYP 119) after Protein Data Bank 1IO7 (Park et al., 2002). The protein is shown as the ribbon. The heme that contains a Fe atom is shown as balls.

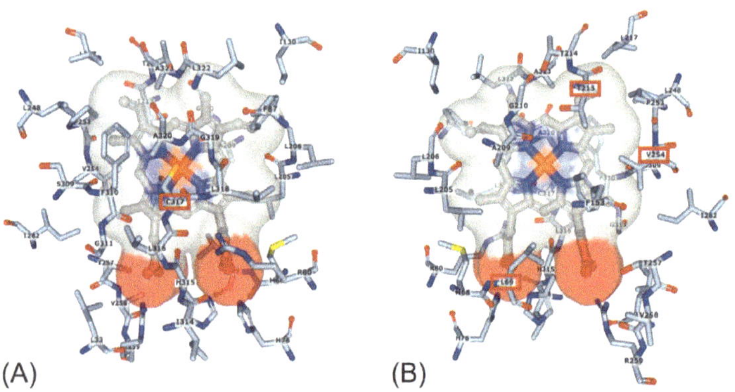

Fig. 3.5.2 Enlarged figures around the heme of wild type (WT)-CYP119. The amino acids reacted with the heme are shown. The number with species of the amino acids is shown, (A) is the same direction as Fig. 2.3.1, and (B) is the reverse of (A). The amino acids in the boxes show the modified amino acids in the study of Dydio et al. (2016). The quadruple mutant CYP-119-Max has G317 in (A)-side, and G213, V69 and L254 in (B)-side.

3.6 EVALUATION OF PROTEINS AS CATALYSTS

The reaction used for evaluation of proteins as catalyst is shown in Fig. 3.6.1. This reaction does not occur with the natural Fe-PIX enzymes. Following the methods summarized in Section 3.3, Dydio et al. (2016) made an Ir(Me)-PIX CYP119 enzyme. This is named as the wild type (WT) Ir (Me)-CYP119 enzyme. However, this protein showed low turnover frequency (TOF) and enantioselectivities (ee) of 0.23 min^{-1} and 0%, respectively. From this WT Ir(Me)-CYP119, they replaced amino acids one by one. They changed C317G [this indicates amino acid mutation of cysteine (C) into glycine (G)] and obtained TOF and ee of 0.13 min^{-1} and 14%, respectively. In both experiments, 5 mM of (**1**) and 0.1 mol% of the catalyst were used.

The mutant of C317G and T213G gave TOF and ee of 9.3 min^{-1} (80-fold to WT) and 68%, respectively. The quadruple mutant C317G, T213G, L69V, and V254L (named CYP119-Max) gave 43 min^{-1} (180-fold) and 94%. The four changed amino acids are shown in boxes in Fig. 3.5.2 (one in the a-side and three in the b-side).

3.7 POSSIBILITY OF METALLOPORPHYRINS WITH INFINITE SELECTIVITY

These studies demonstrate that unknown enzymatic reactivity can be caused by incorporating a metal ion with an accompanying small ligand into a well-known metalloprotein, keeping the native structure of the active site. Selectivity for specific substrates can be achieved readily by directed evolution.

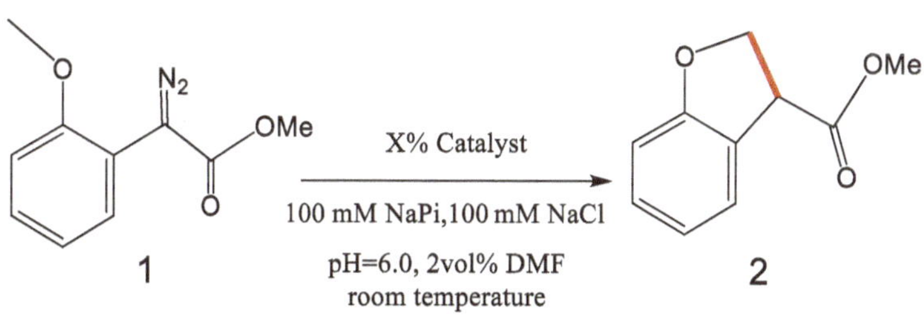

Fig. 3.6.1 The reaction used for evaluation of proteins as catalyst. NaPi is trisodium phosphate. DMF is *N,N*-dimethylformamide [(CH_3) NCHO].

Considering the rich chemistry of free metalloporphyrins and the ease of preparation and evolution of heme proteins containing various metals by the described methods, this methodology should create many new artificial metalloenzymes with diverse and unnatural reactivity. Furthermore, the facile and direct expression of apo-heme proteins could be used in tandem with strategies to incorporate highly active noble-metal complexes of ligands beyond porphyrins. ("Expression" in this context means production of a modified protein by gene manipulation.) Access to a wide range of artificial heme proteins provides almost limitless chances to achieve catalytic reactions with selectivity derived from the interaction of the substrate with a natural binding site.

3.8 INTRODUCTION TO INSERTION OF CARBON-SILICON BONDS BY CYTOCHROME C

Almost 30% of the mass of Earth's crust is made by silicon; however, it is not yet known how carbon-silicon bonds are formed in life. Silicon is abundant, but the synthesizing methods of organosilicon compounds are not so common. Some catalysts containing precious metals, such as Rh, Ir, and Cu, are used in carbene insertion into silanes.

Carbene is a molecule containing a neutral carbon atom with a valence of two, and two unshared valence electrons. The general formula of the carbene is R-(C:)-R. Silane is SiH_4. "Carbene insertion into silanes" means:

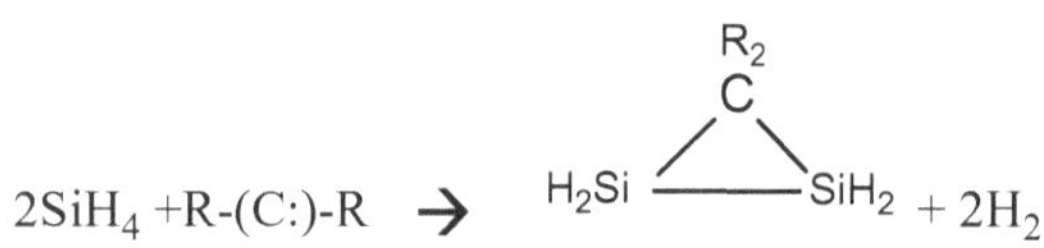

This solution can be used for biocatalysts as unexpected functions to generate catalytic novelty. For example, heme proteins can catalyze a variety of nonnatural carbene-transfer reactions in aqueous media.

3.9 DISCOVERY OF CYTOCHROME C AS THE ENZYME FOR ORGANOSILICON

Kan et al. (2016) discovered that cytochrome c from *Rhodothermus marinus* (Rma cyt c), a Gram-negative, thermohalophilic bacterium from submarine hot springs in Iceland, catalyzed the reaction with 97% enantiomeric excess (ee). This indicated that the reaction took place in an environment where the protein worked excellent stereocontrol.

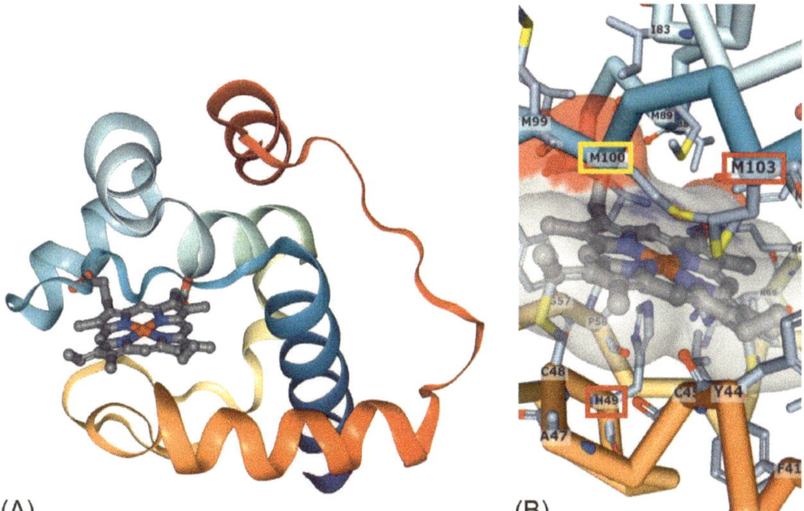

(A) (B)

Fig. 3.9.1 (A) Cytochrome c from wild-type *Rhodothermus marinus* (PDB 3CP5) (Stelter et al., 2008). The bars and balls at the left side indicate the heme. (B) The backbone style of the same protein. The amino acids of H49, M100, and M103 (His49, Met100, and Met103) are indicated by the boxes. The heme is indicated as the hazed balls.

Bacterial cytochromes c are well-studied, functionally conserved electron-transfer proteins that are not known to have any catalytic function in living systems. Other bacterial and eukaryotic cytochrome c proteins also catalyzed the reaction, but with lower selectivities.

The crystal structure of wild-type Rma cyt c (Protein Data Bank: 3CP5, see Fig. 3.9.1; Stelter et al., 2008) reveals that the heme prosthetic group resides in a hydrophobic pocket, with the iron axially coordinated to a proximal His (H49) and a distal Met (M100); the latter is located on a loop. The distal Met, common in cytochrome *c* proteins, is coordinatively labile. Wild-type Rma cyt c became the enzyme (heme) that made Me-EDA (**1**) into the organosilicon compounds (**3**) with high enantiomeric excess (ee) of >97% (see Fig. 3.9.2).

3.10 MUTAGENESIS OF THE WILD TYPE ENZYME

The mutagenesis of triple-mutant V75T (T: threonine), M100D (D: aspartic acid), and M103E (E: glutamic acid) catalyzed the C—Si bond formation in >1500 turnovers and >99% ee. This level of activity is >15 times the total turnovers reported for the best synthetic catalysts for this class of

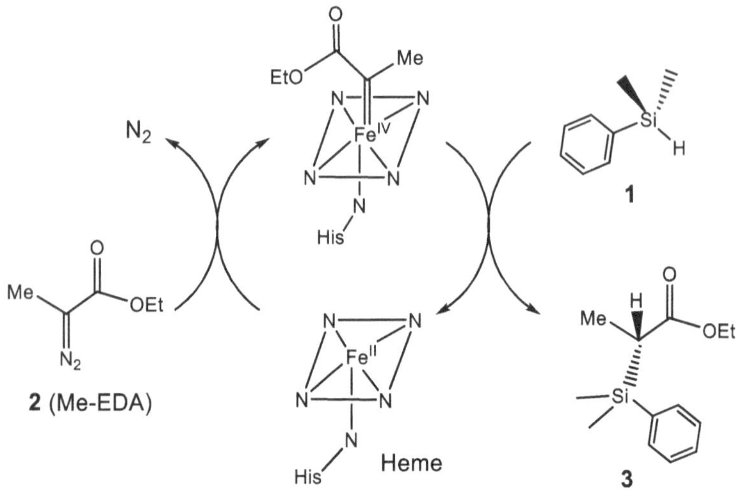

Fig. 3.9.2 Wild-type *R. marinus* cytochrome c became the enzyme (heme) that made Me-EDA (**1**) into the organosilicon compounds (**3**) with high enantiomeric excess (ee) of >97%.

reaction. Relative to the wild-type protein, the evolved triple mutant catalyzes the reaction more than seven times faster, with turnover frequency (TOF) of $46\,\mathrm{min}^{-1}$.

The triple mutant catalyzed the formation of 20 Si-containing products, and most of them are single enantiomers. The reaction includes both electron-rich and electron-deficient silicon reagents, such as ethers, aryl halides, alkyl halides, esters, and amides.

3.11 THE FUTURE OF METALLOENZYMES

The studies shown in this chapter indicate the possibilities for metalloenzymes produced by merging inorganic and organic chemical and biochemical methods. Natural enzymes can be improved by changing metals. For example, transition metals in the enzymes are replaced by abiotic metals such as precious metals or organic complexes with precious metals. In addition, if a protein is used in the metalloenzyme, replacing amino acids based on biochemistry could also be effective. Thus application of both chemical and biochemical strategies is a prerequisite to improve the productivity and selectivity of a metalloenzyme.

REFERENCES

Dydio, P., Key, H.M., Nazarenko, A., Rha, J.Y.-E., Seyedkazemi, V., Clark, D.S., Hartwig, J.F., 2016. An artificial metalloenzyme with the kinetics of native enzymes. Science 354, 102–106.

Kan, S.B.J., Lewis, R.D., Chen, K., Arnold, F.H., 2016. Directed evolution of cytochrome c for carbon-silicon bond formation: bringing silicon to life. Science 354, 1048–1051.

Key, H.M., Dydio, P., Clark, D.S., Hartwig, J.F., 2016. A biological catalysis by artificial haem proteins containing noble metals in place of Iron. Nature 534, 534–537.

Park, S.Y., Yamane, K., Adachi, S., Shiro, Y., Weiss, K.E., Maves, S.A., Sligar, S.G., 2002. Thermophilic cytochrome P450 (CYP119) from Sulfolobus solfataricus: high resolution structure and functional properties. J. Inorg. Biochem. 91, 491–501.

Sigel, A., Sigel, H., Sigel, R.K.O. (Eds.), 2007. The Ubiquitous Roles of Cytochrome P450 Proteins. John Wiley & Sons, West Sussex, UK.

Stelter, M., Melo, A.M., Pereira, M.M., Gomes, C.M., Hreggvidsson, G.O., et al., 2008. A novel type of monoheme cytochrome c: biochemical and structural characterization at 1.23 a resolution of rhodothermus marinus cytochrome c. Biochemist 47, 11953–11963.

Watson, H.C., 1969. The stereochemistry of the protein myoglobin. Prog. Stereochem. 4, 299.

CHAPTER FOUR

Developments of New Biochemical Methods

Contents

4.1 NEW SINGLE-CELL WHOLE-GENOME ANALYSIS METHOD

Rapid advances in DNA sequencing have given us vast amounts of knowledge about genomes of humans by bulk measurements from many cells. However, a single cell of an individual human being has unique genome differences—for example, in each germ cell, somatic cells, etc. The differences include single-nucleotide variations (SNVs), copy-number variations (CNVs), and structural variations. Such genomic changes could cause cancer and other diseases. Thus, characterization of single-cell genomes has attracted increasing attention in recent years (e.g., Gawad et al., 2016). Single-cell genomics becomes more and more important in the case of valuable and rare samples, such as embryonic cells or circulating tumor cells (CTC).

As amounts of genomic DNA in a cell are very few, single-cell genome sequencing highly depends on whole-genome amplification (WGA). In WGA methods, the degenerate oligonucleotide-primed polymerase chain

https://doi.org/10.1016/B978-0-12-817054-0.00004-7
© 2019 Elsevier Inc.
All rights reserved.

reaction (DOP-PCR) is an exponential PCR reaction with degenerate priming (Telenius et al., 1992). Multiple annealing and looping-based amplification cycles (MALBAC) use quasilinear amplification through looping based amplicon protection followed by PCR (Zong et al., 2012). Multiple-displacement amplification (MDA) uses a strand-displacing DNA polymerase to amplify single-stranded DNA exponentially into a hyperbranched structure (Leung et al., 2016). All these methods involve amplification bias and errors.

To reduce such bias and errors, Chen et al. (2017) have developed a new method: Linear Amplification via Transposon Insertion (LIANTI). This combines Tn5 transposition (Goryshin and Reznikoff, 1998) and T7 in vitro transcription (Van Gelder et al., 1990). Random fragmentation and tagging of genomic DNA by Tn5 transposition has been used to prepare DNA sequencing libraries by introducing priming sites for PCR amplification (Adey et al., 2010). However, such exponential amplification has potential amplification bias and errors, limiting its applications in single-cell genomics (Reuter et al., 2016; Zahn et al., 2017). To overcome the error amplification problem, Chen et al. (2017) used the linear amplification.

In LIANTI, genomic DNA from a single cell is randomly fragmented by Tn5 transposition of a specially designed LIANTI transposon including a T7 promoter. The DNA fragments tagged by T7 promoters are linearly amplified into thousands of copies of RNAs through in vitro transcription. The reverse transcription is then followed and the second-strands are synthesized into double-stranded LIANTI amplicons ready for DNA library preparation. LIANTI deletes nonspecific priming and exponential amplification used in other single-cell WGA methods. Therefore, amplification bias and errors are greatly reduced.

Compared with other WGA methods, LIANTI has more uniform amplification and higher sequence coverage. The method enabled the detection of a type of mutation, which involves the gain or loss of regions of the genome. Such detection was hard even using other amplification methods with high resolution. In contrast, Chen et al. (2017) could characterize micro-copy number variations, which are smaller than 100,000 bases, but with a resolution of about 10,000 bases.

4.2 GETTING GPCRs INTO LIPOSOMES

4.2.1 What are GPCRs?

The guanine nucleotide-binding proteins (G proteins) work as molecular switches inside cells, or transduction of signals from outside to inside of the cell, resulting in controlling metabolism or sensing of phenomena. The functions of the G proteins are controlled by hydrolyzing guanosine triphosphate (GTP) to guanosine diphosphate (GDP) (see Fig. 4.2.1). When the G proteins bind to GTP, they are "ON." When they bind to GDP, they are "OFF." The G proteins are a family of enzymes called GTPases.

The G protein-coupled receptors (GPCRs) are known as seven-(pass)-transmembrane domain receptors, 7TM receptors, heptahelical receptors, serpentine receptor, or G protein-linked receptors (GPLR). The GPCRs represent the superfamily of transmembrane receptors found in eukaryotes (Katritch et al., 2013). The example of Human A2a receptor (2YDO) is shown in Fig. 4.2.2A after Lebon et al. (2011). One side of the protein is in a cell, and the other side of the protein is outside of the cell to catch signals from outside (see the Fig. 4.2.2B).

GPCRs recognize a wide variety of structurally diverse agonists (ligands bind to a receptor and activate the receptor) and antagonists (those bind to a receptor and inactivate the receptor) such as hormones, peptides, lipids, nucleotides, and neurotransmitters. In addition to G proteins, GPCRs may bind with multiple intracellular ligands such as arrestin (a small family

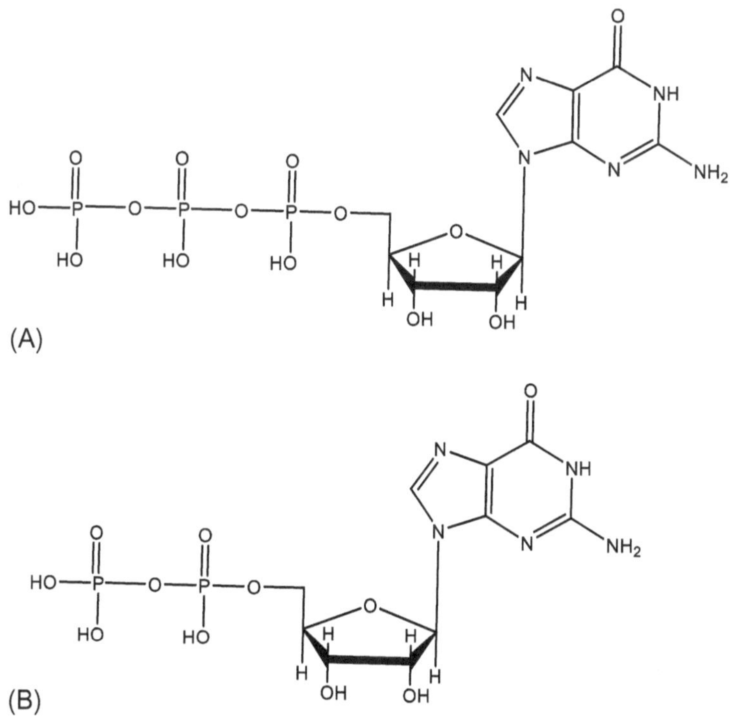

Fig. 4.2.1 (A) Guanosine triphosphate (GTP) and (B) guanosine diphosphate (GDP).

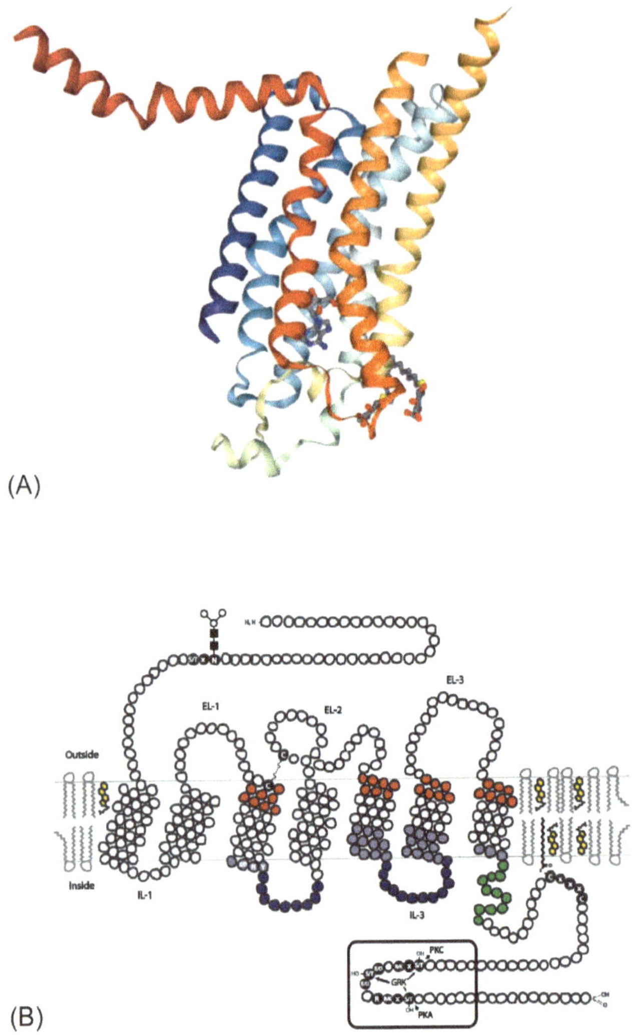

Fig. 4.2.2 (A) Thermostabilised HUMAN A2A receptor with adenosine bound. 2YDO from PDB after Lebon et al. (2011). (B) A two-dimensional diagram of trancemembrane GPCR protein within the plasma membrane localized to a lipid raft. The chain of small circles indicates the protein. The upper and lower sides indicate outside and inside of the cell, respectively. The figure is from Wikipedia: http://upload.wikimedia.org/wikipedia/commons/8/84/GPCR_in_membrane.jpg.

of proteins that regulate GPCRs) and start endocytosis (transporting molecules from outside to inside the cell by transforming cell membrane outside-in).

A class of proteins known as β-arrestins was also found to mediate alternative signaling pathways in response to receptor activation. β-Arrestins, so named because they often oppose the primary activation pathways associated with numerous GPCRs, are now considered to be a critical component of overall GPCR signaling.

4.2.2 Difficulty of GPCR Purification

It is difficult to characterize the precise conditions of signals for GPCR to recognize, particularly, the signaling bias or allosteric effects. One strategy is to separate and purify GPCR, and study the receptor behavior in detergent micelles. Unfortunately, most membrane proteins like GPCRs are unstable in detergent bilayers (micelles) and alter their properties.

To avoid these detergent effects, the membrane proteins are extracted and reconstituted in stable synthetic membranes. Although these methods are effective and do not change the fundamental properties of GPCR function, the methods are very complicated and time-consuming, requiring chromatography to remove detergents.

Furthermore, membrane curvature and composition of cell membranes can change the functions of GPCR. Therefore, it is better to keep the native chemical environment of a whole-cell embedded GPCR.

Recently, Cole et al. (2015) demonstrated the viability of using nonmembrane forming surfactants, such as lysophospholipid (phospholipid of which one of two hydrophobic tails are lacked) analogs and fatty acyl thioesters, as reactive precursors to generate liposomes, and subsequently reconstitute membrane proteins during new phospholipid synthesis.

However, the lysophospholipids used for protein solubilization are modest detergents for the purification and isolation of membrane proteins (Bhamidipati and Hamilton, 1995). Critical micelle concentrations (CMCs) of lysophospholipid are between 4 and 8 μM and low, so that micelles are formed before the surfactant solubilizes a membrane protein (Stangl et al., 2012).

4.2.3 Standard Extraction Method of GPCRs

In the standard extraction and solubilization of GPCRs, nonionic alkyl glucoside detergents, specifically DDM, are used followed by detergent depletion in the presence of 1-palmitoyl-2-oleoyl-*sn*-glycero-3-phospho choline (POPC). The maltoside surfactant induces the micellar-solubilized receptors into fusing with a stable lipid-system. DDM forms oblate ellipsoid micelles, which stabilize GPCRs by encapsulating the bilayer environment, whereas their large micellar size helps in preventing protein-protein aggregation. Furthermore, as DDM has a moderate cmc of 170 μM, it permits higher working concentrations than other conventional detergents such as octyl-β-glucopyranoside (OGP) or *N*,*N*-dimethyldodecylamine-N-oxide (DDAO).

4.2.4 In Situ Lipid Synthesis for Protein Reconstitution Technology (INSYRT)

To overcome these problems, Brea et al. (2017) developed in situ lipid synthesis for protein reconstitution technology (INSYRT), which gives a rapid and selective method for creating GPCR-containing proteoliposomes. The target GPCR was the adenosine A_{2A} receptor ($A_{2A}R$) in synthetic liposomes using new n-dodecyl-*β*-D-maltoside (DDM) thioester analogs.

Based on the optimal characteristics of DDM, Brea et al. (2017) used the DDM-like dodecanoyl maltose thioester, which works as both a protein-solubilizing surfactant and a reactive precursor in their native chemical ligation (NCL) reconstitution method.

Here, INSYRT is described briefly. It was carried out by substituting DDM for the dodecyl maltose thioester to form micellar-solubilized $A_{2A}R$ (see Fig. 4.2.3). The protein-solubilized micelles were reacted with the cysteine-functionalized oleoyl lysophosphatidylcholine through native chemical ligation (NCL). The reaction gave phospholipid, a synthetic analog of native 1-oleoyl-2-palmitoyl-*sn*-glycero-3-phosphocholine (OPPC). Formation of phospholipid leads to stable liposomes with $A_{2A}R$ in the membrane.

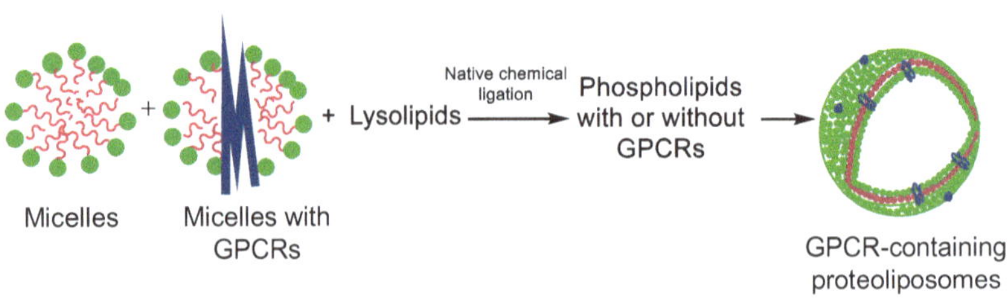

Fig. 4.2.3 Schematic diagram of INSYRT.

The method has the advantages of nonenzymatic and chemo-selective coupling partners (maltose thioester and lysophospholipid), which rapidly react with retaining specificity in buffers. The only byproduct is the eliminated thiomaltose.

The phospholipid made by INSYRT also has an advantage of rapidly constituting the A_{2A} receptor in liposomes with minimum steps. The in situ NCL reaction can perform the concerted depletion oflysophospholipid and the accumulation ofphospholipid in ~20 min. Without postpurifications. The final molar ratio of A_{2A}R:phospholipid was ~1:530.

In summary, INSYRT should be a very powerful biotechnological method in studies of complex membrane proteins or of synthetic cells.

4.3 A FULLY AUTOMATED ACCELERATED PEPTIDE SYNTHESIS MACHINE

The ribosome in *Escherichia coli* can combine eight amino acids per second, each of which is combined with an amide bond or a peptide bond. Researchers require anything from one minute to hours to make only one amide bond in a flask. Mijalis et al. (2017) have speeded this process up by developing a fully automated flow-based solid-phase peptide synthesis equipment. Using this machine, one amide bond can be added in 7 s, and a peptide can be assembled in 40 s per amino acid.

The machine consists of three pumps connected with 50 solutions of amino acids and chemical activators. The reagents are always flowing at a flow rate of 80 mL per minute. The system can make 10–100 times faster than batch-based solid-phase peptide synthesis methods generally used. If the machine were to work nonstop for a year, 0.79 million peptide bond proteins would be synthesized.

4.4 MICROFLUIDIC DEVICE FOR LOCAL STIMULATION OF LYMPH-NODE SLICES

The lymph node is a structurally complex organ of the immune system, whose dynamic cellular arrangements are thought to control much of human health. Currently, no methods exist to stimulate substructures precisely within the lymph node (Fig. 4.4.1A) or analyze local stimulus-response behaviors. This makes it difficult to design therapies rationally for inflammatory disease.

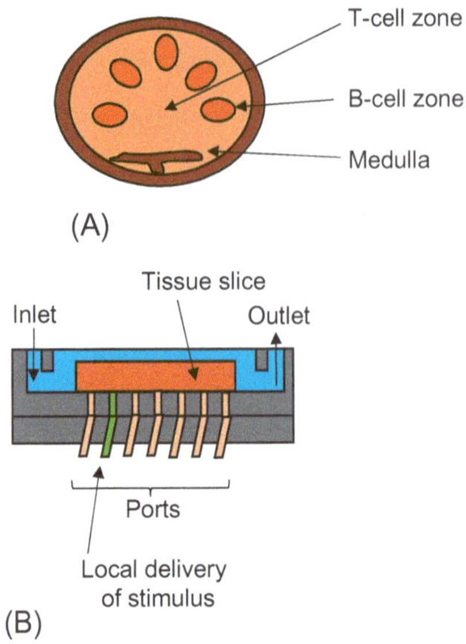

Fig. 4.4.1 (A) Schematic structure of a slice of a limph node. The width is ~2 mm. (B) Side view of the device. The tissue slice is perfused with media. The stimulus solution is delivered through a microchannel to stimulate a local region of the slice.

Ross et al. (2017) succeeded in stimulating locally live lymph node slices with a microfluidic system. Slices maintained the cellular organization of the lymph node by making its core experimentally accessible. The three-layer polydimethylsiloxane device consisted of a perfusion chamber on stimulation ports that are fed by underlying microfluidic channels (see Fig. 4.4.1B).

Fluorescent dextrans similar in size to common proteins (40 and 70 kDa) were delivered to live lymph node slices with $284 \pm 9\,\mu m$ and $202 \pm 15\,\mu m$ spatial resolution, respectively. Five seconds are sufficient to target functional zones of the lymph node.

Delivery of a model therapeutic albumin to specific regions of the lymph node indicated that retention of the drug was larger in the B-cell zone than in the T-cell zone. Together, this work provides a novel platform, the lymph node slice on-a-chip, to target and study local events in the lymph node and to inform the development of new immuno-therapies.

4.5 LIPOSOME PROLIFERATION EXPERIMENT

The Yomo team at Osaka University focused on the sustainable proliferation of phospholipid vesicles. They successfully achieved sustainable

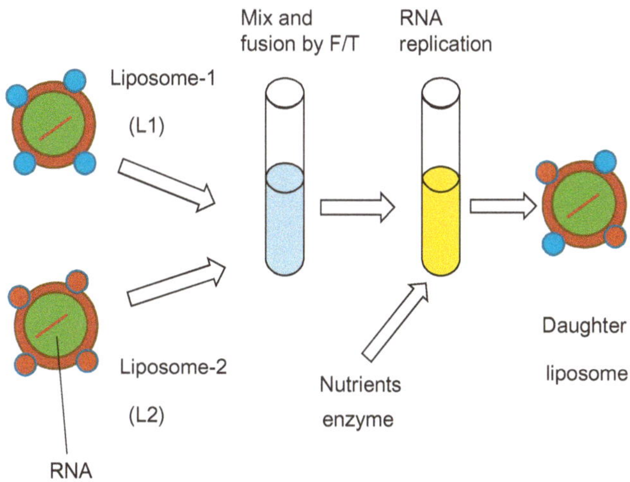

Fig. 4.5.1 The schematic diagram of the cell cycle by Yomo's group (Tsuji et al. 2016).

proliferation of liposomes that replicate RNA within them (Tsuji et al., 2016).

Nutrients for RNA replication and membranes for liposome proliferation were combined by using a modified freeze-thaw technique. The liposomes were mixed, frozen by liquid nitrogen, and centrifuged. The supernatant were then discarded, and dissolved with new buffer solution, because the inner contents of liposomes could be leaked by the freezing. These liposomes showed fusion and fission compatible with RNA replication and distribution to daughter liposomes. A schematic diagram of the experiment is shown in Fig. 4.5.1. The RNAs in daughter liposomes were repeatedly used as templates in the next RNA replication and were distributed to granddaughter liposomes. Liposome proliferation was achieved within 10 cycles.

Therefore, Tsuji et al. (2016) proposed the use of cultivable liposomes as an advanced protocell model, with the implication that the simultaneous supplement of both the membrane material and the nutrients of inner reactions might have enabled protocells to grow sustainably.

REFERENCES

Adey, A., Morrison, H.G., Asan, Xun, X., Kitzman, J.O., Turner, E.H., et al., 2010. Rapid, low-input, low-bias construction of shotgun fragment libraries by high-density **in vitro** transposition. Genome Biol. 11, R119.

Brea, R.J., Cole, C.M., Lyda, B.R., Ye, L., Prosser, R.S., Sunahara, R.K., et al., 2017. In situ reconstitution of the adenosine A_{2A} receptor in spontaneously formed synthetic liposomes. J. Am. Chem. Soc. 139, 3607–3610.

Bhamidipati, S.P., Hamilton, J.A., 1995. Interactions of lyso 1-palmitoylphosphatidylcholine with phospholipids: A 13C and 31P NMR study. Biochemistry 34, 5666–5677.

Chen, C., Xing, D., Tan, L., Li, H., Zhou, G., Huang, L., Xie, X.S., 2017. Single-cell whole-genome analyses by linear amplification via transposon insertion (LIANTI). Science 356, 189–194.

Cole, C.M., Brea, R.J., Kim, Y.H., Hardy, M.D., Yang, J., et al., 2015. Spontaneous reconstitution of functional transmembrane proteins during bioorthogonal phospholipid membrane synthesis. Angew. Chem. Int. Ed. 54, 12738–12742.

Gawad, C., Koh, W., Quake, S.R., 2016. Single-cell genome sequencing: current state of the science. Nat. Rev. Genet. 17, 175–188.

Goryshin, I.Y., Reznikoff, W.S., 1998. Tn5 in vitro transposition. Biol. Chem. 273, 7367–7374.

Katritch, V., Cherezov, V., Stevens, R.C., 2013. Structure-function of the G protein-coupled receptor superfamily. Annu. Rev. Pharmacol. Toxicol. 53, 531–556.

Lebon, G., Warne, T., Edwards, P.C., Bennett, K., Langmead, C.J., et al., 2011. Agonist-bound adenosine A(2a) receptor structures reveal common features of GPCR activation. Nature 474, 521–525.

Leung, K., Klaus, A., Lin, B.K., Laks, E., Biele, J., 2016. Robust high-performance nanoliter-volume single-cell multiple displacement amplification on planar substrates. Proc. Natl. Acad. Sci. U.S.A. 113, 8484–8489.

Mijalis, A.J., Thomas III, D.A., Simon, M.D., Adamo, A., Beaumont, R., et al., 2017. A fully automated flow-based approach for accelerated peptide synthesis. Nat. Chem. Biol. 13, 464–466.

Reuter, J.A., Spacek, D.V., Pai, R.K., Snyder, M.P., 2016. Simul-seq: combined DNA and RNA sequencing for whole-genome and transcriptome profiling. Nat. Methods 13, 953–958.

Ross, A.E., Belanger, M.C., Woodroof, J.F., Pompano, R.R., 2017. Spatially resolved microfluidic stimulation of lymphoid tissue *ex vivo*. Analyst 142, 649–659.

Stangl, M., Veerappan, A., Kroeger, A., Vogel, P., Schneider, D., 2012. Detergent properties influence the stability of the glycophorin a transmembrane helix dimer in lysophosphatidylcholine micelles. Biophys. J. 103, 2455–2464.

Telenius, H., Carter, N.P., Bebb, C.E., Nordenskjöld, M., Ponder, B.A., Tunnacliffe, A., 1992. Degenerate oligonucleotide-primed PCR: General amplification of target DNA by a single degenerate primer. Genomics 13, 718–725.

Tsuji, G., Fujii, S., Sunami, T., and Yomo, T. 2016. Sustainable proliferation of liposomes compatible with inner RNA replication. Proc. Natl. Acad. Sci. U.S.A. 113: 590–595.

Van Gelder, R.N., von Zastrow, M.E., Yool, A., Dement, W.C., Barchas, J.D., Eberwine, J.H., 1990. Amplified RNA synthesized from limited quantities of heterogeneous cDNA. Proc. Natl. Acad. Sci. U.S.A. 87, 1663–1667.

Zahn, H., Steif, A., Laks, E., Eirew, P., VanInsberghe, M., et al., 2017. Scalable whole-genome single-cell library preparation without preamplification. Nat. Methods 14, 167–173.

Zong, C., Lu, S., Chapman, A.R., Xie, X.S., 2012. Genome-wide detection of single-nucleotide and copy-number variations of a single human cell. Science 338, 1622–1626.

CHAPTER FIVE

Topics of Bioluminescence and Chemoluminescence

Contents

5.1 MECHANISM OF GLOWING MUSHROOMS

Many living organisms emit light, which is called bioluminescence. In many cases, the chemical oxidation of a luciferin substrate catalyzed by a luciferase enzyme emits light. The luciferin reacts with molecular oxygen, and becomes a high-energy intermediate (HEI), whose decomposition produces the oxyluciferin when in an excited state. This excited metabolite emits fluorescent light, resulting in visible light as bioluminescence (Shimomura, 2012; Wilson and Hastings, 2013).

The luciferin/luciferase systems of several bioluminescent insects, bacteria, and marine animals have been characterized to work by the corresponding oxyluciferins. This knowledge has yielded major technological advances, such as genetically encoded labels using fluorescent proteins, etc.

Bioluminescent fungi are spread throughout the world (Oliveira et al., 2012). Recently, Kaskova et al. (2017) solved the molecular mechanism and reasoning behind fungal bioluminescence. Furthermore, they identified and characterized the fungal oxyluciferin based on the study of Purtov et al. (2015).

Biochemistry for Materials Science
https://doi.org/10.1016/B978-0-12-817054-0.00005-9
© 2019 Elsevier Inc.
All rights reserved.

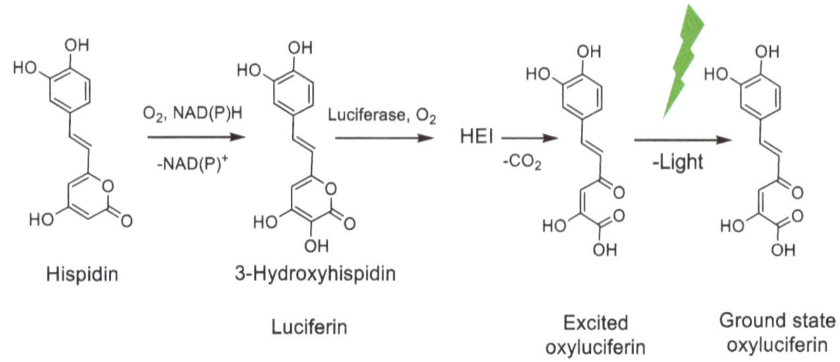

Fig. 5.1.1 General mechanism of fungal bioluminescence.

Kaskova et al. (2017) reported the chemical structure of fungal luciferin and its biosynthetic precursor. The enzymatic oxidation of the precursor (hispidin) by an NAD(P)H (nicotinamide adenine dinucleotide phosphate)-dependent enzyme, hispidin-3-hydroxylase, produces 3-hydroxyhispidin, which acts as the fungal luciferin (Fig.5.1.1).

5.2 FIRST NATURAL FLUORESCENCE IN FROGS

Fluorescence is the re-emission of longer wavelength lights after the absorption of shorter wavelength lights. This phenomenon is broadly found in living organisms, which are distantly related in both marine and terrestrial environments.

In aquatic vertebrates, fluorescence occurs in cartilaginous and ray-finned fishes, and sea turtles. In terrestrial vertebrates, it is known to occur only in parrots. Amphibians (frogs, toads, salamanders, newts, and geckos) have various skin coloration caused by an integumental pigmentary system. The different types of chromatophore cells create various colors. Although the chromophores have been studied, no fluorescence in any of the 7600 species of amphibians has been reported.

Taboada et al. (2017) first found fluorescence in the amphibian group in the tree frog *Hypsiboas punctatus*. They showed that fluorescence is produced by a combination of lymph and glandular emissions, with pigmentary cell filtering in the skin (see Fig. 5.2.1).

The chemical origin of fluorescence was found to be based on fluorescent compounds derived from dihydroisoquinolinone, which are named hyloins (Hyloin-L1, Hyloin-L2 and Hyloin-G1; see Figs. 5.2.2A–C, respectively).

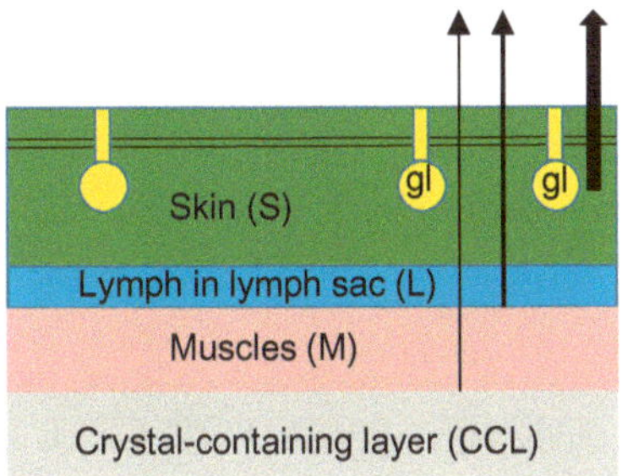

Fig. 5.2.1 Anatomy of fluorescence in *H. punctatus*. The thickness of the arrow corresponds to the intensity of the fluorescence emission. Each tissue layer attenuates the emission. The fluorescence mainly comes from green skin with lymph. "gl" indicates yellow glandular secretions. In CCL, guanine crystals are contained.

(A)

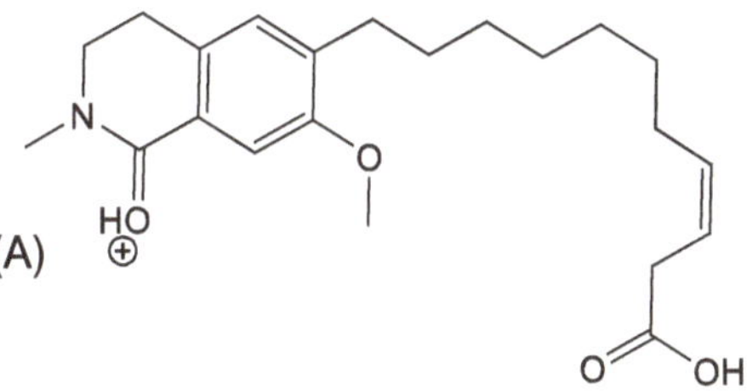

(B)

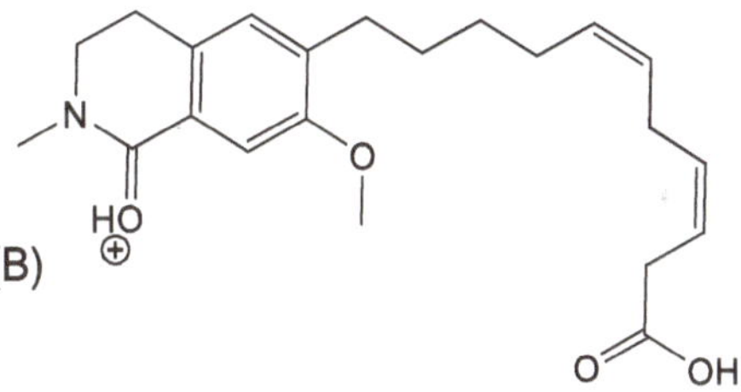

(C)

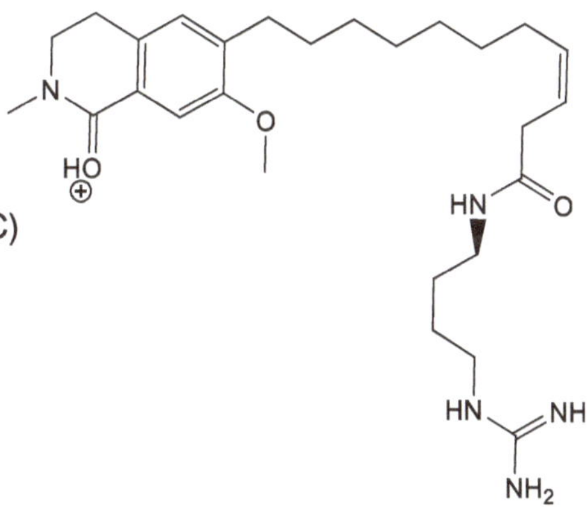

Fig. 5.2.2 Fluorescent compounds of the *H. punctatus*. (A) Hyloin-L1. (B) Hyloin-L2. (C) Hyloin-G1.

The frog appears yellow in full daylight, but lime green in twilight conditions by contribution of the fluorescence. The enhanced brightness of the frog matches the sensitivity of night vision in amphibians. Taboada et al. suggest that the fluorescence is a more widespread phenomenon than previously thought, although its purpose has not yet been explained.

5.3 CHEMOLUMINESCENCE FOR CELL IMAGING

5.3.1 Introduction

The chemoluminescence probe is the most sensitive method for determination of enzyme concentration, because it has a very high signal-to-noise ratio. Therefore, the chemoluminescence probe is used in immunoassays including DNA. Most chemoluminescence probes emit lights by the mechanism as follows: the reaction starts by oxidation of an oxidizing agent, which is oxidized into an unstable peroxide, making a light-emission-material into its excited state; it then rapidly decays to the ground state with emitting light. The oxidation-based mechanism is utilized for the activation of common chemoluminescence substrates such as luminol or oxalate esters. Furthermore, oxidation-activated chemoluminescence has been used to detect and image reactive oxygen species (ROS) in vitro and in vivo.

5.3.2 Schaap's Chemoluminescence Probe

Schaap's adamantylidene – dioxetane (see Fig. 5.3.1A, structure I) is the only known chemoluminescence probe containing a stable dioxetane (Schaap et al., 1987a,b,c). This probe does not require an oxidation step to trigger its chemoluminescence, meaning that the probe can be applied to a wide range of chemical and biological activities. Schaap's dioxetane I has an analyte-responsive protecting group (PG), which masks the phenol moiety of the probe. Removal of this PG forms an unstable phenolate – dioxetane species II, which decomposes through a chemoexcitation process to produce the excited intermediate benzoate ester III and adamantanone. The excited intermediate decays to its ground state through the emission of a blue photon of 470 nm.

5.3.3 Modified Chemoluminescence Probe by Hananya et al. (2016)

In order to obtain brighter chemoluminescence, the excited species must be a more efficient emitter. This also means that the excited species must have a

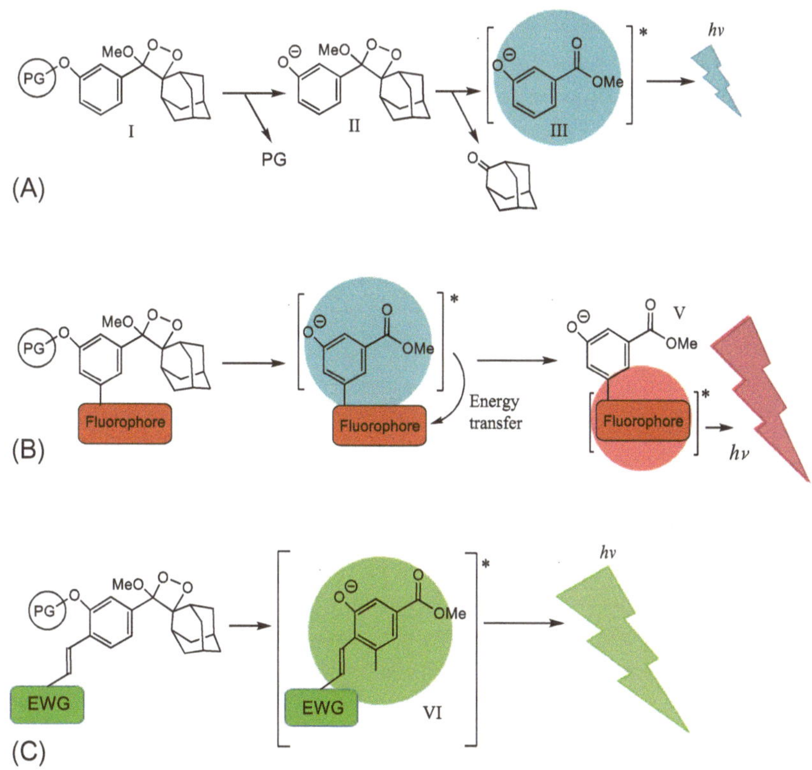

Fig. 5.3.1 (A) Schaap's adamantylidene-dioxetane reactions. (B) The enhancement of chemiluminescent light by Hananya et al. (2016). (C) The new study by Green et al. (2017). EWG indicates electron withdrawing group, such as acrylate and acrylonitrile. The color and size of the "lightning figure" correspond to the approximate color and relative intensity of chemiluminescence.

higher fluorescence quantum yield. Benzoate ester III emits strong light in organic solvents like dimethyl sulfoxide (DMSO; $(CH_3)_2SO$), but it is a weak fluorophore in aqueous conditions. Therefore, the chemoluminescence generated by the dioxetane probe in water is very weak.

Hananya et al. (2016) have recently reported a practical synthetic route to adamantylidene – dioxetane conjugated with fluorescent dyes. The dioxetane – fluorophore conjugate decomposes on its activation to generate the typical excited benzoate, which can transfer its energy to the attached fluorophore, giving a highly emissive intermediate (Fig. 5.3.1B, compound V). The chemiluminescent light drastically increases under physiological conditions.

5.3.4 Brightest Chemoluminescence Probe by Green et al. (2017)

Green et al. (2017) followed a methodology that is also based on the incorporation of a substituent on the benzoate species obtained during the chemoexcitation pathway of Schaap's adamantylidene – dioxetane probe.

When electron-withdrawing groups (EWG; such as -CN) were installed (see Fig. 5.3.1C), a drastic substituent effect on the chemoluminescence efficiency was observed. The best quantum yield of chemoluminescence was ~530 nm and more than three orders of magnitude higher than that of a standard, commercially available adamantylidene – dioxetane probe. These are the most powerful chemoluminescence dioxetane probes synthesized so far. Furthermore, they are suitable for use even in aqueous conditions at pH = 7.4.

One of these probes developed by Green et al. (2017) was capable of providing high-quality chemoluminescence cell images based on endogenous activity of β-galactosidase. This is the first demonstration of cell imaging achieved by a nonluciferin small-molecule probe with direct chemoluminescence. The method presented here should enable efficient chemoluminescence probes for various biochemical applications in sensing and imaging.

REFERENCES

Green, O., Eilon, T., Hananya, N., Gutkin, S., Bauer, C.R., et al., 2017. Opening a gateway for chemiluminescence cell imaging: distinctive methodology for design of bright chemiluminescent dioxetane probes. ACS Cent. Sci. 3, 349–358.

Hananya, N., Eldar Boock, A., Bauer, C.R., Satchi-Fainaro, R., Shabat, D., 2016. Remarkable enhancement of chemiluminescent signal by dioxetane – fluorophore conjugates: Turn-on chemiluminescence probes with color modulation for sensing and imaging. J. Am. Chem. Soc. 138, 13438–13446.

Kaskova, Z.M., Dörr, F.A., Petushkov, V.N., Purtov, K.V., Tsarkova, A.S., et al., 2017. Mechanism and color modulation of fungal bioluminescence. Sci. Adv. 3, e1602847.

Oliveira, A.G., Desjardin, D.E., Perry, B.A., Stevani, C.V., 2012. Evidence that a single bioluminescent system is shared by all known bioluminescent fungal lineages. Photochem. Photobiol. Sci. 11, 848–852.

Purtov, K.V., Petushkov, V.N., Baranov, M.S., Mineev, K.S., Rodionova, N.S., et al., 2015. The chemical basis of fungal bioluminescence. Angew. Chem. Int. Ed. 54, 8124–8128.

Schaap, A.P., Chen, T.-S., Handley, R.S., DeSilva, R., Giri, B.P., 1987a. Chemical and enzymatic triggering of 1,2-dioxetanes. 2: fluorideinduced chemiluminescence from tert-butyldimethylsilyloxy-substituted dioxetanes. Tetrahedron Lett. 28, 1155.

Schaap, A.P., Handley, R.S., Giri, B.P., 1987b. Chemical and enzymatic triggering of 1,2-dioxetanes. 1: aryl esterase-catalyzed chemiluminescence from a naphthyl acetate-substituted dioxetane. Tetrahedron Lett. 28, 935–938.

Schaap, A.P., Sandison, M.D., Handley, R.S., 1987c. Chemical and enzymatic triggering of 1,2-dioxetanes. 3: alkaline phosphatase-catalyzed chemiluminescence from an aryl phosphate-substituted dioxetane. Tetrahedron Lett. 28, 1159–1162.

Shimomura, O., 2012. Bioluminescence: Chemical Principles and Methods. World Scientific.

Taboada, C., Brunetti, A.E., Pedron, F.N., Neto, F.C., Estrin, D.A., et al., 2017. Naturally occurring fluorescence in frogs. PNAS 114, 3672–3677.

Wilson, T., Hastings, J.W., 2013. Bioluminescence: Living Lights, Lights for Living. Harvard Univ. Press, London, United Kingdom.

FURTHER READING

Hou, J., Wang, L., Zhang, P., Xu, Y., Ding, L., 2015. Facile synthesis of carbon dots in an immiscible system with excitation-independent emission and thermally activated delayed fluorescence. Chem. Commun. 51, 17768–17771.

Jacoby, M., 2017. Copper dimers reduce NOx. C&EN 95, 8. August 28.

Liu, J., Wang, N., Yu, Y., Yan, Y., Zhang, H., et al., 2017. Carbon dots in zeolites: A new class of thermally activated delayed fluorescence materials with ultralong lifetimes. Sci. Adv. 3, e1603171.

Olah, G.A., Molnár, Á., 1995. Hydrocarbon Chemistry. Wiley & Sons, New York.

Paolucci, C., Khurana, I., Parekh, A.A., Li, S., Shih, A.J., et al., 2017. Dynamic multinuclear sites formed by mobilized copper ions in NOx selective catalytic reduction. Science. https://doi.org/10.1126/science.aan5630.

Webster-Gardiner, M.S., Chen, J., Vaughan, B.A., McKeown, B.A., Schinski, W., et al., 2017. Catalytic synthesis of "super" linear alkenyl arenes using an easily prepared Rh(I) catalyst. J. Am. Chem. Soc. 139, 5474–5480.

Zhu, S., Meng, Q., Wang, L., Zhang, J., Song, Y., et al., 2013. Highly photoluminescent carbon dots for multicolor patterning, sensors, and bioimaging. Angew. Chem. Int. Ed. 52, 3953–3957.

CHAPTER SIX

New Antibiotic and Antivirus Agents, and Knowledge

Contents

6.1 ANTIBIOTIC FROM CATERPILLAR GUT BACTERIA

6.1.1 Introduction

Insects are the largest group of animals on Earth. They are the most important herbivores in the world. Herbivorous insects eat and use large amounts of plant material. In their eating behavior, they ingest many kinds of harmful microorganisms from their food, which contains abundant microbial flora, or from the surrounding environment. The limited resources of the insects can compete against such invading microbes. The microbes also increase the risk of infection via the gut, which is probably the main route of insect infection in nature. Although the insects are attacked repeatedly by such microorganisms, herbivorous insects are resistant to microbial infections and flourish even in microbe-rich environments.

Therefore it should hardly be astonishing that they have evolved effective antimicrobial strategies, such as the immune system or the physical and chemical barriers in the gut. Lepidopteran larvae have the strongly alkaline pH in their gut. The indigenous microbiota or the native gut microbiota are acknowledged to be the part of the host defense system against microbial invaders in many animals.

Biochemistry for Materials Science
https://doi.org/10.1016/B978-0-12-817054-0.00006-0
© 2019 Elsevier Inc.
All rights reserved.

For example, the microbiota of bumblebees protects against an intestine parasite *Crithidia bombi*. A normal microbial community of mice can mediate pathogens from the gut lumen in primary infections with *Salmonella typhimurium*. Despite accumulating observations that the native gut microbiota are involved in combating infections, the species achieving these protective effects has not been either identified from the microbial consortia, or cultured outside the host. In addition, the molecular mechanisms are poorly understood.

6.1.2 Antimicrobial Activity of the Cotton Leafworm

Shao et al. (2017) targeted the antimicrobial activity of a cultural gut symbiotic bacterium in the cotton leafworm *Spodoptera littoralis* (*Lepidoptera Noctuidae*). The worm represents one of the most polyphagous and widespread insect herbivores in temperate regions, causing substantial economic losses in crop production. Like most lepidopteran larvae, cotton leafworms have a simple alimentary canal, like a tube without any special structure. However, a large bacterial population—more than 10^7 colony-forming units (CFU)/mL—is observed in the gut. Recent studies of the taxonomic composition of this gut microbiota reveal a simple community of *Enterococcus* and *Clostridium spp*. In particular, *Enterococcus mundtii* (*E. mundtii*) is found in normal individuals regardless of diet or generation.

Although *E. mundtii* is a major constituent of the gut microbiota, other ubiquitous environmental enterococci, such as *E. faecalis* and *E. casseliflavus*, often started out at high density in early generation when *E. mundtii* had not yet become abundant in the gut. *Enterococcus* bacteria occur in plants, soil, and water, but also have adaptability to the gastrointestinal tracts of animals resulting in serious infections. In particular, *E. faecalis* and *E. casseliflavus* have many mechanisms to invade the host, whereas *E. mundtii* rarely becomes a pathogen.

As *enterococci* strains with various virulence factors have appeared in many habitats, these pathogens that are orally acquired by *S. littoralis* larvae from their diet (eggshell and plants) and the surrounding environment are a critical challenge to the host. However, it is well known that they are quickly depleted from the microbiota with an increase of the *E. mundtii* population, which stabilizes in the gut together with clostridia as core components of the flora.

It seems likely that the gut-resident *E. mundtii* has an inherent advantage, and they have already blocked the gut niche at an early stage. This potentially confers invasion resistance benefit on *E. mundtii*'s host. These observations encouraged the research team to isolate and identify the enterococci in *S. littoralis*, and to investigate whether *E. mundtii* specifically inhibit nonindigenous invading microbes, and if so, by what mechanism.

Finally, the team found that the dominant symbiotic bacterium *Enterococcus mundtii* in the general herbivore *S. littoralis* secretes a stable class IIa bacteriocin (mundticin KS) against invading bacteria, but not against other gut residents, facilitating the normal development of host gut microbiota (see Fig. 6.1.1). The crystal structure of mundticin (2ZRR in PDB) is shown in Fig. 6.1.2A and mundticin KS is shown in Fig. 6.1.2B. A mundticin-defective strain lost inhibitory activity. Furthermore, purified mundticin has been proven to cure infected larvae. Thus, the constitutively produced antimicrobials by native extracellular symbionts create a significant chemical barrier in the gut.

This unique property also benefits *E. mundtii* itself by providing a competitive advantage, contributing to its dominance within complex microbial settings and its prevalence of *S. littoralis*. It is also likely to promote long-term cooperative symbiosis between them.

In addition, this mundticin KS has a possibility to be a new food preservative. This peptides can be a new antimicrobial food additives or coatings for food contact surfaces. It is very interesting that the new antibiotics appear to be hiding in the gut microbiomes of various creatures.

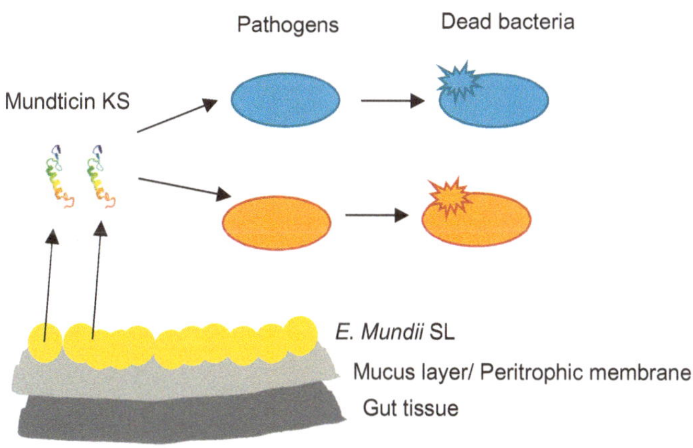

Fig. 6.1.1 Schematic diagram of the antimicrobial system in the gut of *Spodoptera littoralis*.

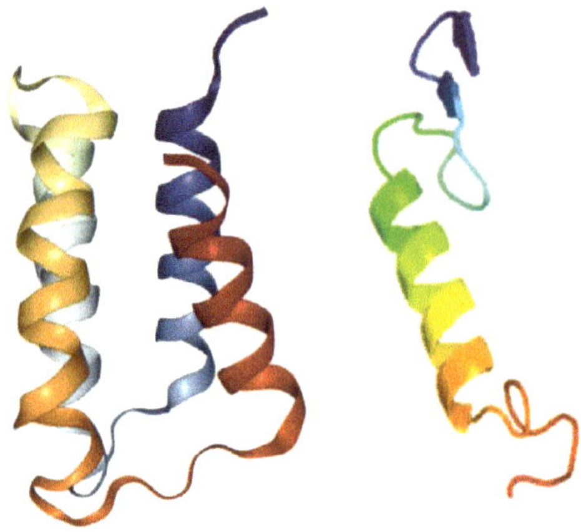

Fig. 6.1.2 Crystal structure of an immunity protein which contributes to the self-protection of bacteriocin-producing Enterococcus mundtii 15-1A: (A) mundticin (2ZRR in PDB) by Jeon et al. (2009); (B) mundticin KS by Shao et al. (2017).

6.2 NOVEL ANTIBIOTIC FROM FUNGAL IN TOXIC LAKE

Microorganisms from isolated environments are a good source of new and bioactive compounds. However, the extremophilic fungi isolated from the acidic, metal-rich waters of Berkeley Pit Lake did not show little evidence for antibiotic production in the preliminary study.

Stierle et al. (2017) performed a careful co-fermentation of *Penicillium fuscum* and *P. camembertii/clavigerum*, and found eight new 16-membered-ring macrolides, berkeleylactones A–H, as well as the known antibiotic macrolide A26771B, patulin, and citrinin.

There was no evidence of the production of the berkeleylactones or A26771B by either fungus when grown as axenic cultures.

The structures were deduced from analyses of spectral data, and the absolute configurations of berkeleylactones A and F (see Fig. 6.2.1A and B) were determined by single-crystal X-ray crystallography.

Berkeleylactone A exhibited the most effective antimicrobial activity of the macrolide series, with low micromolar activity (MIC = 1–2 μg/mL) against four MRSA strains, as well as *Bacillus anthracis*, *Streptococcus pyogenes*, *Candida albicans*, and *Candida glabrata*. Berkeleylactone A did not inhibit protein synthesis nor target the ribosome, unlike other macrolide antibiotics,

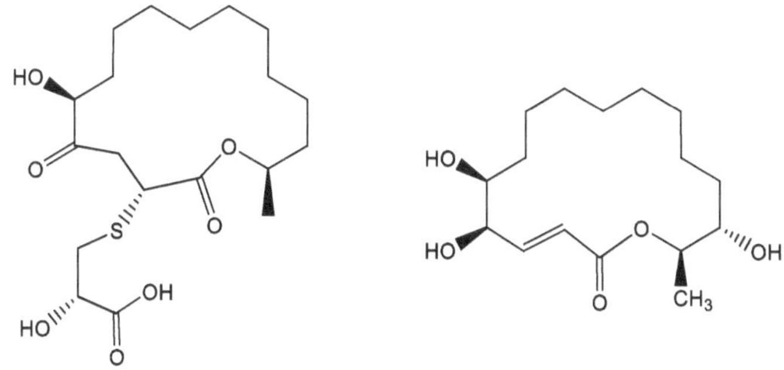

(A) (B)

Fig. 6.2.1 (A) Berkeleylactone A. (B) Berkeleylactone F.

from the mode of action studies. This suggests a new mode of action for its antibiotic activity.

6.3 FLU-KILLING PEPTIDES WERE FOUND FROM FROG MUCUS

Antiviral treatment is the first line of defense during pandemics, since although vaccines give protection against influenza A viruses, there is not enough time to produce vaccines. Present antiviral drugs are affected by drug resistance; therefore, developing new antivirals is essential.

Holthausen et al. (2017) studied peptides from the skin of the South Indian frog *Hydrophylax bahuvistara* and found that one of these, which they named "urumin" and which was made of 27 amino acids, is virucidal for H1 hemagglutinin-bearing human influenza A viruses (H1HA). This peptide specifically targeted the conserved stalk region of H1 hemagglutinin and was effective against drug-resistant H1 influenza viruses. It also protected naive mice from lethal influenza infection.

Urumin is a unique antiinfluenza virucide that targets the hemagglutinin stalk region, similar to targeting of antibodies induced by universal influenza vaccines. Thus urumin has the potential to be the first-line antiviral treatments during influenza pandemics.

6.4 WHY ZIKA VIRUS TARGETS CEREBRAL NEURAL PRECURSORS

Zika virus, which broke out in Brazil, causes a serious disease called neonatal microcephaly. Zika targets cerebral neural precursors, which are a cell population essential for cortical development. However, the reason for this neurotropism is unclear.

Chavali et al. (2017) have found that the neural RNA-binding protein Musashi-1 (MSI1) interacts with the Zika genome and enables viral replication. The Zika infection disturbs the binding of MSI1 to its intrinsic targets, so that neural stem cell function is disturbed by deregulated expression of factors.

Chavali et al. (2017) further showed that MSI1 is highly expressed in neural progenitors of the human embryonic brain and mutates in individuals with autosomal recessive primary microcephaly (MCPH). Selective MSI1 expression in neural precursors could therefore explain the exceptional weakness of these cells to the Zika infection.

6.5 A NEW ANTIMICROBIAL AGENT: DRGN-1 BASED ON KOMODO DRAGONS

Chung et al. (2017) have identified a histone H1-derived peptide from the Komodo dragon (*Varanus komodoensis*), called VK25. Based on this peptide, they designed a synthetic peptide called DRGN-1, of which the sequence is shown below. (Abbreviations of amino acids are shown in Table 2.1.1.)

VK25 SPKKTKPVKPKKVA

DRGN-1 PSKKTKPVKPKKVA

Chung et al. evaluated the antimicrobial and antibiofilm activity of both peptides against *Pseudomonas aeruginosa* and *Staphylococcus aureus*. DRGN-1 exhibited more potent antimicrobial and antibiofilm activity than VK25, and permeabilized bacterial membranes. Wound healing was significantly enhanced by DRGN-1 in both uninfected and mixed biofilm (*Pseudomonas aeruginosa* and *Staphylococcus aureus*)-infected murine wounds.

In a scratch wound closure assay used to elucidate the wound healing mechanism, the peptide promoted the migration of HEKa keratinocyte cells, which was inhibited by mitomycin C (proliferation inhibitor) and AG1478 (epidermal growth factor receptor inhibitor). DRGN-1 also activated the EGFR-STAT1/3 pathway.

Thus, DRGN-1 is a candidate for use as a topical wound treatment. The new peptide DRGN-1 exhibits pathogen-directed and host-directed activities in promoting the clearance and healing of polymicrobial (*P. aeruginosa* and *S. aureus*) biofilm-infected wounds. The effectiveness of this peptide cannot be attributed solely to its ability to act upon the bacteria and disrupt the biofilm; it also reflects the peptide's ability to promote keratinocyte migration. When DRGN-1 was applied in a murine model, infected wounds treated with it healed significantly faster than did untreated wounds, or wounds treated with other peptides. The host-directed mechanism of action was via the EGFR-STAT1/3 pathway. The pathogen-directed mechanism of action was via antibiofilm activity and antibacterial activity through membrane permeabilization.

This new peptide may have potential as a future therapeutic tool for treating infected wounds.

REFERENCES

Chavali, P.L., Stojic, L., Meredith, L.W., Joseph, N., Nahorski, M.S., et al., 2017. Neurodevelopmental protein Musashi 1 interacts with the Zika genome and promotes viral replication. Science 357, 83–88.

Chung, E.M.C., Dean, S.N., Propst, C.N., Bishop, B.M., van Hoek, M.L., 2017. Komodo dragon-inspired synthetic peptide DRGN-1 promotes wound-healing of a mixed-biofilm infected wound. Biofilms Microbiomes. 3. https://doi.org/10.1038/s41522-017-0017-2.

Holthausen, D.J., Lee, S.H., Kumar, V.T.V., Bouvier, N.M., Krammer, F., et al., 2017. An amphibian host defense peptide is virucidal for human H1 hemagglutinin-bearing influenza viruses. Immunity 46, 587–595.

Jeon, H.J., Noda, M., Matoba, Y., Kumagai, T., Sugiyama, M., 2009. Crystal structure and mutagenic analysis of a bacteriocin immunity protein, Mun-im. Biochem. Biophys. Res. Commun. 378, 574–578.

Shao, Y., Chen, B., Sun, C., Ishida, K., Hertweck, C., Boland, W., 2017. Symbiont-derived antimicrobials contribute to the control of the lepidopteran gut microbiota. Cell Chem. Biol. 24, 66–75.

Stierle, A.A., Stierle, D.B., Decato, D., Priestley, N.D., Alverson, J.B., et al., 2017. The Berkeleylactones, antibiotic macrolides from fungal coculture. J. Nat. Prod. 80, 1150–1160.

CHAPTER SEVEN

Chirality in Space and Brains

Contents

7.1 CHIRAL IMAGE

Fig. 7.1.1 shows two types of amino acids. The chemical composition is the same, but two amino acids cannot be overlapped, because two images are mirror images. This is called a chiral image, and either of two is called enantiomers. These are called D-amino or L-amino acids. For amino acids, D or L can be identified by the "CORN" rule.

Practically, COOH, R, and NH_2 are arranged around the chiral center carbon with H away from the viewer. When CO→R→N is counter-clockwise, it is the L form. If the rotation is clockwise, it is the D form. All amino acids synthesized or used by living things are L-amino acids, with a few exceptions. For example, there are many D-amino acids in brains with high D/L ratios (see Section 7.5).

Biochemistry for Materials Science
https://doi.org/10.1016/B978-0-12-817054-0.00007-2

© 2019 Elsevier Inc.
All rights reserved.

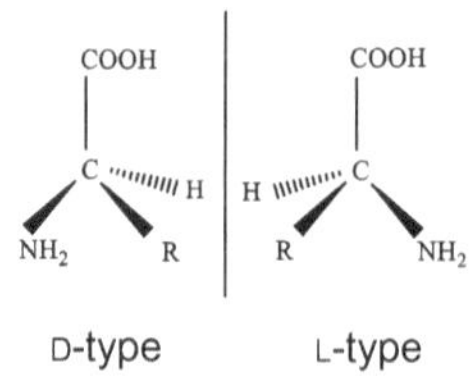

Fig. 7.1.1 Isomers of amino acids. The left and right configurations are D- and L-amino acids, respectively.

7.2 CHIRALITY IN METEORITES

The majority of biological sugars and their derivatives are the D-mirror image rather than the L-mirror image. Nucleic acids on Earth are composed of only D-sugars.

The Murchison meteorite (CM2) is a famous carbonaceous meteorite that has a high concentration of carbon. Pizzarello et al. (2003) separated isovaline (see Fig. 7.2.1A) from many stones of the Murchison meteorite, and measured enantiomer excess (ee %) by GC-MS. They found that L-isovaline are enriched in the range of 0–15.2 ee%. The proteins in space should be racemic; therefore, the origin of such chirality has been unknown.

In contrast, Pizzarello et al. (2012) found D-*allo*-isoleucine (see Fig. 7.2.1B) in the Renazzo-type carbonaceous chondrites (CR) revealed up to 60 ee %. This was attributed to the cosmochemical alteration processes.

Cooper and Rios (2016) analyzed the chirality of straight-chained mono sugar acids (aldonic acids) of three carbons (3C) to six carbons (6C) in the Murchison and Murray (CM2) carbonaceous chondrites (see Fig. 7.2.2 for structures of aldoses and ketoses). They found the D-enantiomer excess in

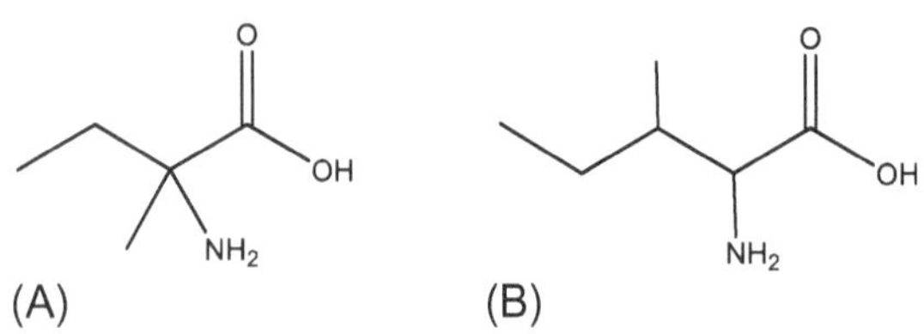

Fig. 7.2.1 (A) Isovaline and (B) allo-isoleucine.

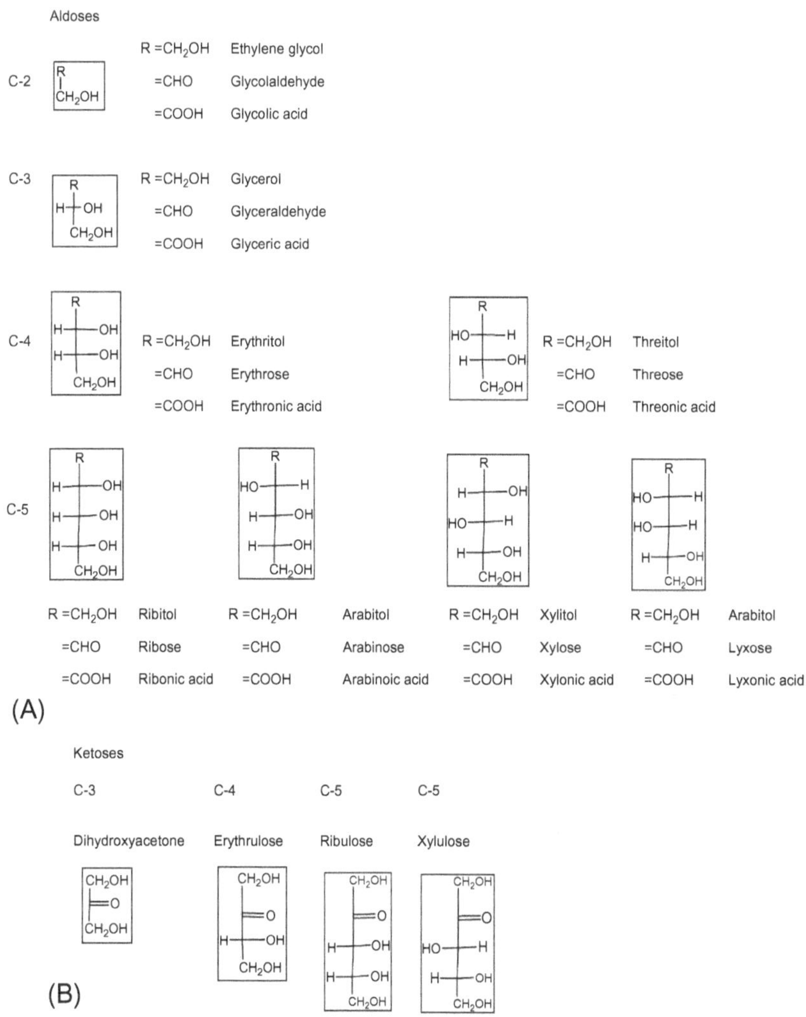

Fig. 7.2.2 Structures and names of linear and branched (A) aldoses and (B) ketoses. Only D-enantiomer forms in the Fischer projections are indicated.

the aldonic acids, which increase with increasing carbon number. In contrast, other sugar acids and sugar alcohols are racemic. Furthermore, only D-ribose, D-2′-deoxyribose, and D-glucose were found in nucleic acids and polysaccharides. Therefore, it is suggested that early meteoritic compounds may have influenced the enantiomer enrichment of subsequent biological sugars and their derivatives in early lives on Earth.

7.3 THE CHALLENGE FOR DIRECT MEASUREMENT OF CHIRALITY OF ORGANIC MATERIALS ON A COMET BY A SPACECRAFT

In order to know directly the chirality of organic materials on the solar system, the spacecraft Rosetta was launched by a European Space Agency (ESA) spacecraft. Rosetta (see Fig. 7.3.1A) targeted a coma (nucleus) of the comet 67P/Churyumov_Gerasimenko (abbreviated as 67PCG). Rosetta was launched in 2004, and the spacecraft reached 67PCG in 2014. It successfully approached the coma of 67PCG, and sent photographs of 67PCG (see Fig. 7.3.2).

One of the special points of the Rosetta project is that it had a lander on the coma of 67PCG attached on Rosetta named Philae (see Fig. 7.3.1B). This had a mass spectrometer to measure enantiomer enrichment (Myrgorodska et al., 2015). It was detached from Rosetta in 2014, and

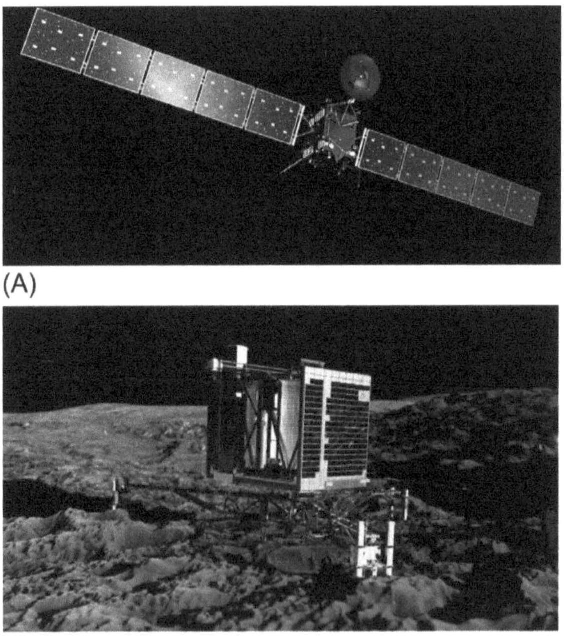

Fig. 7.3.1 Artist's impression of (A) Rosetta and (B) Philae. Rosetta is a spacecraft targeted a coma of the comet 67P/Churyumov_Gerasimenko (67PCG). Philae is a lander on the coma of 67PCG. *Image Credit: (A) ESA/ATG medialab. https://www.flickr.com/photos/europeanspaceagency/11206647984 (B) ESA. http://sci.esa.int/jump.cfm?oid=53559.*

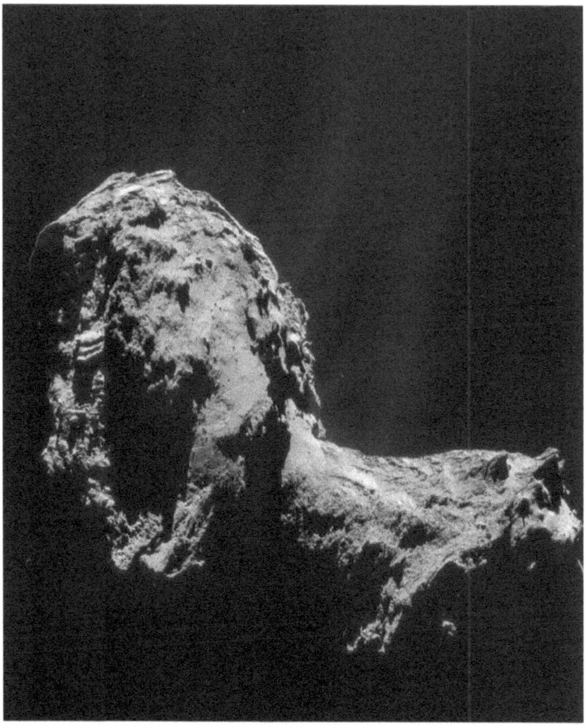

Fig. 7.3.2 A photograph of the comet 67P/Churyumov_Gerasimenko. It was taken from 31 km from the center of the comet. The resolution is 3 m/pixel. The jet activities of water and carbon dioxide gases are seen. As the jet activities are stronger in the center than other locations, the comet became a dumbbell like figure. The surface seems to be made of hard organic materials and ice. The inside should be porous and amorphous ice. "Deep fried ice cream" is a good analogy. *Image Credit: ESA/Rosetta/NAVCAM. http://www.nasa.gov/sites/default/files/thumbnails/image/pia18899.jpg.*

landed on the surface of 67PCG. The first touchdown was successful; however, Philae rebounded on the surface of 67PCG several times, and seemed to be settled in the dark cleavage of the coma where the sunshine was not sufficient to generate the energy required to communicate with Rosetta.

7.4 DISCOVERY OF CHIRAL MOLECULE IN INTERSTELLAR MOLECULE

McGuire et al. (2016) discovered the interstellar chiral molecule propylene oxide (CH_3CHCH_2O), shown in Fig. 7.4.1. The chiral molecule was detected by the radio astronomy toward the Sagittarius B2 North

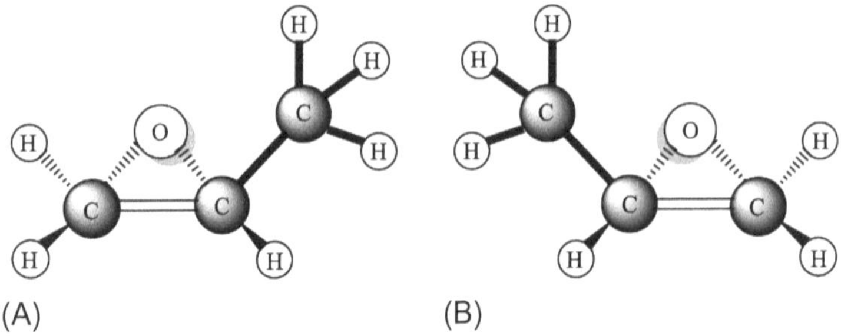

Fig. 7.4.1 The molecular structure of (A) S-propylene oxide and (B) R-propylene oxide. Spheres mean hydrogen (H), carbon (C) and oxygen (O) atoms.

[SgrB2(N)] molecular cloud, the preeminent source for new complex-molecular detections in the interstellar medium (ISM). The propylene oxide was observed using data from the publicity available Prebiotic Interstellar Molecular (PRIMOS) project at the Green Bank Telescope (GBT), which gives high-resolution, high-sensitivity spectral survey data toward SgrB2 (N) from 1 to 50 GHz. The chirality was separated by the radio telescope data of a rotational excitation temperature of ~5 K at molecular density of $N_T = 1 \times 10^{13}\,\mathrm{cm}^{-2}$ and a velocity of $64\,\mathrm{km\,s}^{-1}$. However, by this method, it cannot be determined which molecule in Fig. 7.4.1 is enriched.

7.5 HIGH ABUNDANCES OF D-AMINO ACIDS IN MOUSE BRAINS

7.5.1 Discovery of D-amino Acids in Invertebrates

Amino acids are the most important molecules in nature. Asparagine was the first discovered amino acid. Amino acids were then found from protein hydrolysates. In 1851, Louis Pasteur found that most common amino acids have optical activity caused by their differing orientations around the α-carbon. The initial discovery was that only L-amino acids were found in nature, and D-amino acids were laboratory artifacts (Corrigan, 1969; Cline, 1997).

The idea that D-amino acids are naturally existent or biologically related appeared in the report of Stevens et al. (1951). It was the first report that D-amino acids, especially D-alanine and D-glutamic acids, were related to biological activity. Existence of D-amino acids in living systems then began to be considered common. In 1969, a review with 30 examples of D-amino acids in invertebrates was published (Corrigan, 1969). However, in many cases, biological roles were unknown.

7.5.2 Discovery of D-Amino Acids in Vertebrates (Human and Animal Tissues)

The appearance of new bioanalytical techniques made it possible to isolate and identify D-amino acids far easily in a greater variety of biological samples, especially of vertebrates, by the end of the 20th century (e.g., Dunlop et al., 1986; Armstrong et al., 1993a). In 1986, free D-aspartic acid was found in human and animal tissues (Dunlop et al., 1986). Subsequently, free L- and D-amino acids were reported in pathologically relevant human urine, plasma, cerebrospinal fluid, and amniotic fluid.

Armstrong et al. (1993b) found that a nonproteinic amino acid, D-pipecolic acid (see Fig. 7.5.1A), was an indicator of the severity of a neurological genetic disease. Janecka et al. (2004) showed that D-amino acid-containing peptides had special functions binding to specific opiate receptors. There are some reports of free D-amino acids in various mammalian tissues. Mothet et al. (2000) found that D-serine (Fig. 7.2.1E) was a coagonist of the *N*-methyl-D-aspartate (NMDA)-type glutamate receptor (NMDA is shown in Fig. 7.2.1C), which could occupy the glycine binding site (see Fig. 7.5.2). Free D-serine was separated from the mammalian forebrain, where the highest concentrations for NMDA receptors are found (Hashimoto et al., 1992, 1993; Nagata et al., 1994).

Hartman et al. (2015) applied D-leucine as an effective treatment for seizures of epilepsia in mice, but the mechanism was unclear. D-Serine and D-leucine are two recent and prominent examples of D-amino acids found in brain tissues having specific function or actions. Most D-amino acids are metabolized by either D-amino acid oxidase (DAO) or D-aspartic acid oxidase. D-amino acid oxidase (DAO) or D-aspartic acid oxidase. D-Amino acid oxidase is important for regulating levels of select D-amino acids and is crucial to help control levels of D-serine.

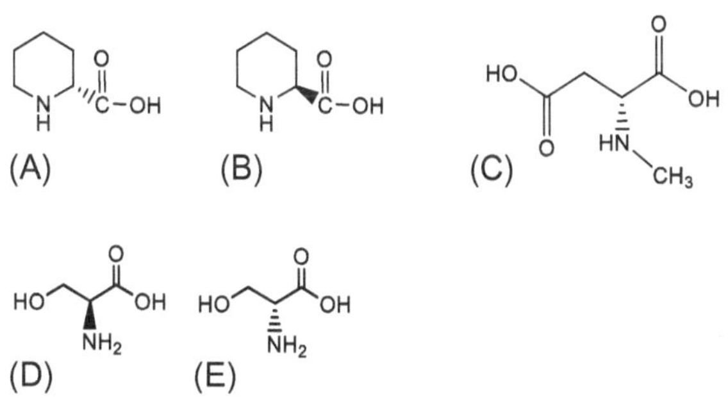

Fig. 7.5.1 Pipecolic acids, *N*-methyl-D-aspartate (NMDA) and serines. (A) D-pipecolic acid. (B) L-pipecolic acid. (C) *N*-methyl-D-aspartate (NMDA). (D) L-serine. (E) D-serine.

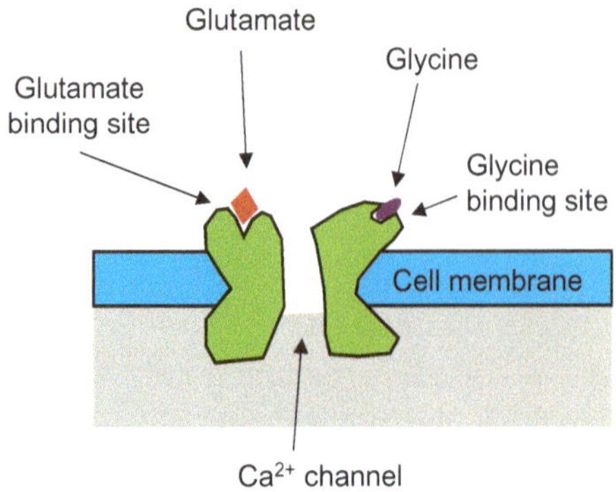

Fig. 7.5.2 NMDA receptor. The NMDA (*N*-methyl-D-aspartate) -type receptor was activated by glutamate receptor.

Although D-amino acids should have the unique biological roles, fundamental studies on normal, baseline levels of most L- and D-amino acids in the brain and blood of any mammalian entity are limited. In addition, sensitivity to detect D-amino acids was insufficient (Hamase et al., 2007, 2010).

7.5.3 Determination of L- and D-Amino Acids in Hippocampus, Cortex Regions in Brain and Blood

Hippocampus (roughly the central part of a brain) and cortex (the outermost layer of a brain) regions are the most common epilepsy sources in the brain (Hartman et al., 2015). Determining baseline levels of D-amino acids in these regions will support attempts to establish essential normative data that further studies can use in finding functions of neurological pathology and presently unknown D-amino acid processes.

Weatherly et al. (2017) gave the most complete characterization of brain and blood amino acid levels in mice for the first time.

Transcardial perfusion is a standard technique involving vascular perfusion of anesthetized animals for preservation of tissues for analysis (Gage et al., 2012). It is unclear how much of each amino acid is in the intravascular and extravascular spaces, which could give useful information about their potential physiological roles. Regarding D-amino acid analyses, perfusion has been applied only to the determination of D-serine and its function in the activity of DAO in the human central nervous system (Sasabe et al., 2014). The effect of perfusion in the analysis of a spectrum of L- and D-amino acids in brain tissues has never been examined. Weatherly et al. (2017) measured D-amino acids in both perfused and nonperfused brain tissues for the first time.

7.5.4 Concentrations of D- and Total Amino Acids in Cortex, Hippocampus, and Blood in Mice

Table 7.5.1 shows the averages of amino acid data for the perfused cortex, hippocampus, and blood of mice ($n=7$, two data from one mouse were averaged).

When concentrations of D-amino acids in perfused tissues were plotted against those in nonperfused tissues, there were similar slopes in both cortex and hippocampus, indicating systematic effects of the perfusion, but no systematic effects of the perfusion in the concentrations. Therefore, in Table 7.5.1, data of the perfused tissues are used.

The three amino acids of glutamic acid, glutamine, and aspartic acid are the highest in brain tissues. Concentration levels of D-amino acids are also shown in Table 7.5.1 (abbreviations of amino acids are shown in Table 2.1.1). Concentrations of D-glutamine, D-aspartic acid, and D-serine have the highest values. It was unexpected that no detectable amount of D-glutamic acid was found, because the concentration of total glutamic acid was the highest among all amino acids. Concentrations of D-amino acid levels were always lower than the corresponding L-amino acid levels. Concentrations of D-glutamine in the nonperfused cortex and that of D-isoleucine in the nonperfused hippocampus are approximately four times less than those of corresponding total amino acids. Concentrations of D-phenylalanine, D-leucine, D-isoleucine, and D-allo-isoleucine were almost the lowest.

7.5.5 Concentrations of Total Amino Acids vs *D*% of Cortex, Hippocampus, and Blood in Mice

Concentrations (μg/mg) of total amino acids vs *D*% of the cortex, hippocampus, and blood in mice are shown in Fig. 7.5.3. There are weak inverse relationships between the concentrations of total amino acids and *D*% for the cortex and hippocampus.

There are two interesting features in these data. First, glutamic acid (Glu) is the most abundant of all free amino acids; however, it has almost the lowest levels in *D*%. This may indicate that D-glutamic acid metabolism exists. Second, it has been observed that the least abundant amino acid is *allo*-isoleucine (Allo-Ile), and has almost equal amounts of D- and L-isomers in the hippocampus. Third, high *D*% of serine (D-serine) was observed in both the cortex and hippocampus.

D-Serine works in the NMDA-type glutamate receptor. D-Amino acid oxidase (DAO) affects only D-serine, but D-serine racemase, a reversible

Table 7.5.1 Total Amino Acid and D-Amino Acid Levels and *D*% in Cortex, Hippocampus and Blood of Perfused Mice. The Data is From Weatherly et al. (2017)

	Cortex			Hippocampus			Blood		
	Total µg/mg	D-Amino acid µg/mg	*D*%	Total µg/mg	D-Amino acid µg/mg	*D*%	Total µg/mg	D-Amino acid µg/mg	*D*%
Leu	0.04	0.006	16.4	0.04	0.009	20.4	0.04	0.0001	0.3
Ser	0.15	0.047	32	0.17	0.053	30.4	0.01	0.001	5.8
Ala	0.25	0.025	9.8	0.26	0.034	13	0.05	0.00005	0.09
Asp	0.79	0.106	13.5	0.65	0.064	9.8	0.19	0.04	18.9
Thr	0.05	0.007	11.3	0.08	0.007	8.4	0.1	0.002	2
Gln	3.34	<0.00001	<0.1	2.96	<0.00001	<0.1	0.19	<0.00001	<0.01
Val	0.02	0.003	12.3	0.05	0.009	20.9	0.03	0.0003	1.1
Asn	0.03	0.007	23.3	0.05	0.014	25.6	0.01	0.0004	3.6
Glu	1.19	0.048	4.1	1.11	0.058	5.2	0.12	0.00005	0.04
Ile	0.01	0.001	14.8	0.01	0.002	24.2	0.01	0.0001	0.8
Allo-Ile	0.002	0.0005	23.4	0.002	0.001	50.1	0.0004	0.0001	26.1
Phe	0.03	0.005	19.1	0.03	0.008	27.3	0.02	0.003	15.6

Notes: Each data is the average of 14 samples of 7 mice. $D\% = D/(D+L) \times 100\%$.

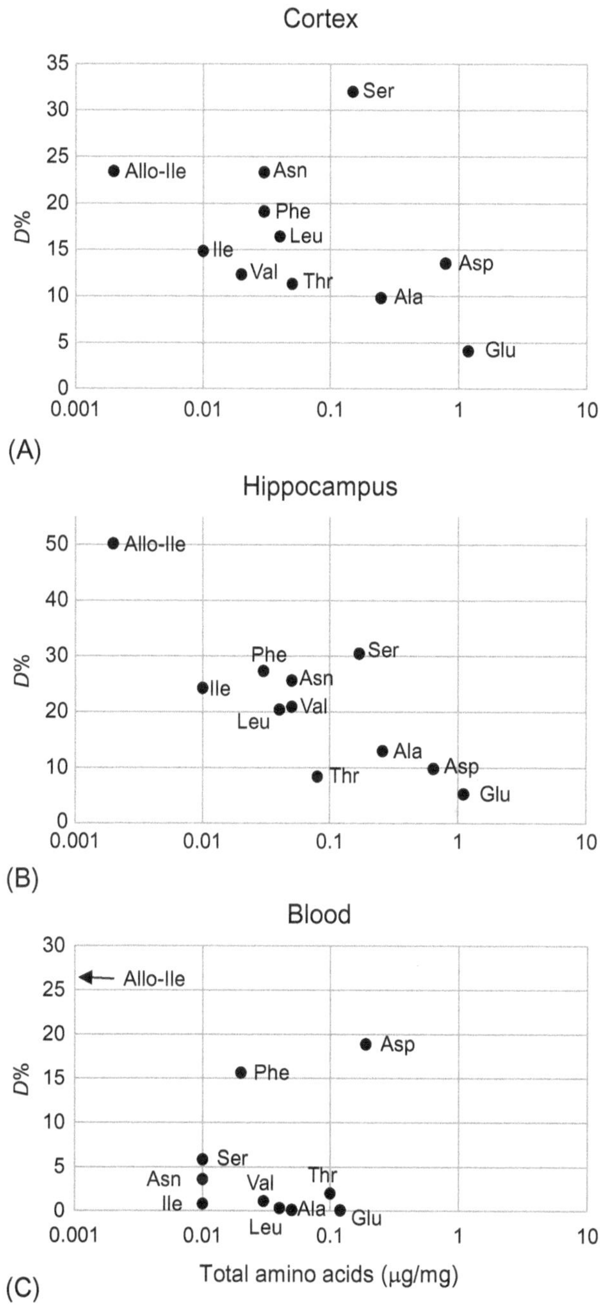

Fig. 7.5.3 Concentrations of total amino acid (μg/mg) vs *D*% of (A) cortex. (B) hippocampus and (C) blood in mice.

enzyme, affects both L- and D-isomers of serine. Formation of D-serine from L-serine by serine racemase should be an important mechanism for maintaining the levels of D-serine in rat brain (Wolosker et al., 1999). The *D*% values should be a good indicator for serine racemase activity. Currently, serine and aspartate racemases are the only D-amino acid racemases reported in mammalian brains (Wolosker et al., 1999; Kim et al., 2010). The results of Weatherly et al. (2017) suggest that there might be other D-amino acid racemases in mammalian brains that cause the high *D*% for phenylalanine (Phe), asparagine (Asn), isoleucine (Ile), etc.

7.6 CONCLUSIONS

The study of Weatherly et al. (2017) has given the most complete and the latest baseline analysis of L- and D-amino acids in mouse brain tissues. There is little difference in the total amino acid levels in the hippocampus and cortex, but all D-amino acid levels in the cortex were ~13% lower than those in the hippocampus. D-Aspartic acid is exceptionally twice as high as that in the cortex and has higher blood levels than all other D-amino acids, making it more available to all tissues. Results suggest that other processes exist beyond DAO and racemases that control broad levels of D-amino acids in brain tissues. Total amino acid concentration vs *D*% shows an inverse relationship. This suggests that there might be other D-amino acid racemases in mammalian brains that keep the high % D-amino acid values. Branched chain amino acids' (BCAAs, such as leucine, valine, isoleucine, and alloisoleucine) levels in blood are similar compared to hippocampus and cortex tissues, but D-amino acid levels for BCAAs are significantly higher in the hippocampus and cortex tissues than in blood. It should be noted that glutamic acid, which is the most predominant free amino acid, had no measurable level of its D-isomer. It is possible that the D-glutamate metabolism in the brain is a unidirectional process and not a cycle, for which reason and mechanism future studies are required.

REFERENCES

Armstrong, D.W., Gasper, M., Lee, S.H., Zukowski, J., Ercal, N., 1993a. D-Amino acid levels in human physiological fluids. Chirality 5, 375–378.

Armstrong, D.W., Zukowski, J., Ercal, N., Gasper, M., 1993b. Stereochemistry of pipecolic acid found in the urine and plasma of subjects with peroxisomal deficiencies. J. Pharm. Biomed. Anal. 11, 881–886.

Cline, D.B., 1997. On the determination of the physical origin of homochirality in life. Comments Nucl. Part. Phys. 22, 131–154.
Cooper, G., Rios, A.C., 2016. Enantiomer excesses of rare and common sugar derivatives in carbonaceous meteorites. PNAS 113, E3322–E3331.
Corrigan, J.J., 1969. D-Amino acids in animals. Science 164, 142–149.
Dunlop, D.S., Neidle, A., McHale, D., Dunlop, D.M., Lajtha, A., 1986. The presence of free D-aspartic acid in rodents and man. Biochem. Biophys. Res. Commun. 141, 27–32.
Gage, G.J., Kipke, D.R., Shain, W., 2012. Whole animal perfusion fixation for rodents. J. Vis. Exp. 3564.
Hamase, K., Morikawa, A., Ohgusu, T., Lindner, W., Zaitsu, K., 2007. Comprehensive analysis of branched aliphatic D-amino acids in mammals using an integrated multi-loop two-dimensional column-switching high-performance liquid chromatographic system combining reversed-phase and enantioselective columns. J. Chromatogr. A 1143, 105–111.
Hamase, K., Miyoshi, Y., Ueno, K., Han, H., Hirano, J., et al., 2010. Simultaneous determination of hydrophilic amino acid enantiomers in mammalian tissues and physiological fluids applying a fully automated micro-two-dimensional high-performance liquid chromatographic concept. J. Chromatogr. A 1217, 1056–1062.
Hartman, A.L., Santos, P., O'Riordan, K.J., Stafstrom, C.E., Hardwick, J.M., 2015. Potent anti-seizure effects of D-leucine. Neurobiol. Dis. 82, 46–53.
Hashimoto, A., Nishikawa, T., Hayashi, T., Fujii, N., Harada, K., Oka, T., Takahashi, K., 1992. The presence of free D-serine in rat brain. FEBS Lett. 296, 33–36.
Hashimoto, A., Kumashiro, S., Nishikawa, T., Oka, T., Takahashi, K., et al., 1993. Embryonic development and postnatal changes in free Daspartate and D-serine in the human prefrontal cortex. J. Neurochem. 61, 348–351.
Janecka, A., Fichna, J., Janecki, T., 2004. Opioid receptors and their ligands. Curr. Top. Med. Chem. 4, 1–17.
Kim, P.M., Duan, X., Huang, A.S., Liu, C.Y., Ming, G.-L., Song, H., Snyder, S.H., 2010. Aspartate racemase, generating neuronal D-aspartate, regulates adult neurogenesis. Proc. Natl. Acad. Sci. U.S.A. 107, 3175–3179.
McGuire, B.A., Carroll, P.B., Loomis, R.A., et al., 2016. Discovery of the interstellar chiral molecule propylene oxide (CH_3CHCH_2O). Science 352, 1449–1452.
Mothet, J.-P., Parent, A.T., Wolosker, H., Brady Jr., R.O., Linden, D.J., Ferris, C.D., Rogawski, M.A., Snyder, S.H., 2000. D-Serine is an endogenous ligand for the glycine site of the N-methyl-D-aspartate receptor. Proc. Natl. Acad. Sci. U.S.A. 97, 4926–4931.
Myrgorodska, I., Meinert, C., Martins, Z., d'Hendecourt, L.L.S., Meierhenrich, U.J., 2015. Molecular chirality in meteorites and interstellar ices, and the chirality experiment on board the ESA cometary Rosetta mission. Angew. Chem. Int. Ed. 54, 1402–1412.
Nagata, Y., Horiike, K., Maeda, T., 1994. Distribution of free D-serine in vertebrate brains. Brain Res. 634, 291–295.
Pizzarello, S., Zolensky, M., Turk, K.A., 2003. Nonracemic isovaline in the Murchison meteorite: chiral distribution and mineral association. Geochim. Cosmochim. Acta 67, 1589–1595.
Pizzarello, S., Schrader, D.L., Monroe, A.A., Lauretta, D.S., 2012. Large enantiomeric excesses in primitive meteorites and the diverse effects of water in cosmochemical evolution. PNAS 109, 11949–11954.
Sasabe, J., Suzuki, M., Imanishi, N., Aiso, S., 2014. Activity of D-amino acid oxidase is widespread in the human central nervous system. Front. Synaptic Neurosci. 6, 14.
Stevens, C.M., Halpern, P.E., Gigger, R.P., 1951. Occurrence of D-amino acids in some natural materials. J. Biol. Chem. 190, 705–710.

Weatherly, C.A., Du, S., Parpia, C., Santos, P.T., Hartman, A.L., Armstrong, D.W., 2017. D-Amino acid levels in perfused mouse brain tissue and blood: a comparative study. ACS Chem. Neurosci. 8, 1251–1261.
Wolosker, H., Blackshaw, S., Snyder, S.H., 1999. Serine racemase: a glial enzyme synthesizing D-serine to regulate glutamate N-methyl-D-aspartate neurotransmission. Proc. Natl. Acad. Sci. U.S.A. 96, 13409–13414.

FURTHER READING

Greenstein, J.P., Birnbaum, S.M., Otey, M.C., 1953. Optical and enzymatic characterization of amino acids. J. Biol. Chem. 204, 307–321.

CHAPTER EIGHT

Bacteria-Related Materials and Topics

Contents

8.1 INFLAMMATION ENCOURAGES BACTERIOPHAGE TRANSFER BETWEEN *SALMONELLA* SPP.

Bacteriophages sometimes encode virulence factors of bacteria, and are also important driving forces for bacterial pathogen evolution, such as *Salmonella enterica Typhimurium* (S.Tm). S.Tm genomes occur in several prophages. Phage transfer (lysogenic conversion) is a key driving force for genomic diversification between closely related enterobacteriaceae. Extensive studies for the phage transfer have been done in vitro. Phage transfer is observed in hosts infected by *Staphylococcus aureus*, *Escherichia coli*, *Pseudomonas aeruginosa*, or *Streptococcus pyogenes*. However, little is known about factors controlling phage transfer in vivo.

In order to study the SopEΦ transfer dynamics in vivo, Diard et al. (2017) used SL1344 as the phage donor. A murine *Salmonella Typhimurium* (*S*.Tm) diarrhea model was used to study the transfer of SopEΦ, a prophage from *S*. Tm SL1344, to *S*.Tm ATCC14028S (see Fig. 8.1.1). Diard et al. (2017) found that gut inflammation and enteric disease triggered >55% of lysogenic conversion of ATCC14028S within 3 days. Without inflammation, SopEΦ transfer was reduced to 10^{-5} times. This was because inflammation (e.g., reactive oxygen species, reactive nitrogen species, hypochlorite) triggers SOS response of the bacteria and encourages expression of the phage antirepressor Tum, resulting in enhancing free phage production and

Biochemistry for Materials Science
https://doi.org/10.1016/B978-0-12-817054-0.00008-4

© 2019 Elsevier Inc.
All rights reserved.

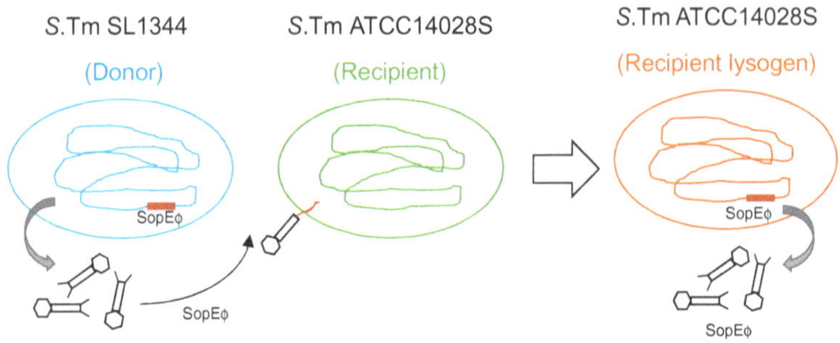

Fig. 8.1.1 Phage transfer experiment. In SL1344 (donor), the SopEΦ prophage was inserted. ATCC14028S was used as the recipient, which does not hold any bacteriophages. Phage transfer occurred in vivo, and recipient lysogen formed.

subsequent transfer. Mucosal vaccination reduced intestinal S.Tm population and consequent inflammation, thereby halting SopEΦ transfer. Diard et al. (2017) concluded that vaccination may be a general method for stopping pathogen evolution that requires disease-driven phage transfer.

8.2 STRUCTURE OF SUPERLONG ICE-BINDING PROTEIN OF THE ANTARCTICA BACTERIUM

Bacterial adhesins are modular cell-surface proteins that mediate adherence to other cells, surfaces, and ligands.

The Antarctic bacterium *Marinomonas primoryensis* uses a 1.5-MDa adhesin (*M. primoryensis* ice-binding protein, MpIBP; see Fig. 8.2.1) comprising over 130 domains to position it on ice at the top of the water column for better access to oxygen and nutrients. Guo et al. (2017) have reconstructed this 0.6-μm-long adhesin and have established complementary roles for its five distinct regions.

Domains in region I (RI) tether the adhesin to the outside of the bacterium and pass it through the outer membrane.

RII comprises ~120 identical immunoglobulinlike β-sandwich domains that rigidify on binding Ca^{2+} to project the adhesion regions RIII and RIV.

RIII contains ligand-binding domains that join diatoms and bacteria together in a mixed-species community on the underside of sea ice where incident light is maximal.

RIV is the ice-binding domain, and the terminal RV domain contains several "repeats-in-toxin" motifs and a noncleavable signal sequence that target proteins for export via the type I secretion system.

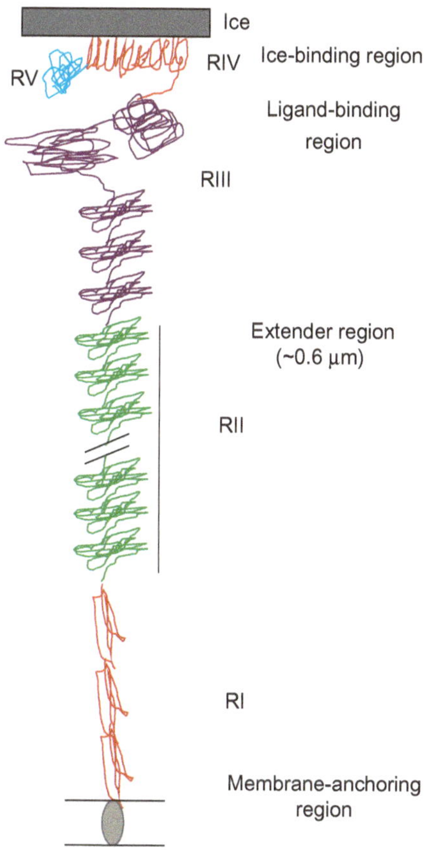

Fig. 8.2.1 Structure of superlong ice-binding protein of the Antarctic bacterium (MpIBP).

A similar structural architecture is present in the adhesins of many pathogenic bacteria, and offers a guide to finding and blocking binding domains to weaken infectivity.

8.3 THE APPEARANCE OF MITOCHONDRIA

The most important evolution was the appearance of mitochondria. The mitochondria are thought to have formed when a bacteria cell began to live inside an archaeal cell as a form of endosymbiosis (a mutually beneficial relationship in which one organism lives inside another). The bacterium is considered to have provided its host cell with additional energy, and the interaction eventually resulted in a eukaryotic cell, which has the

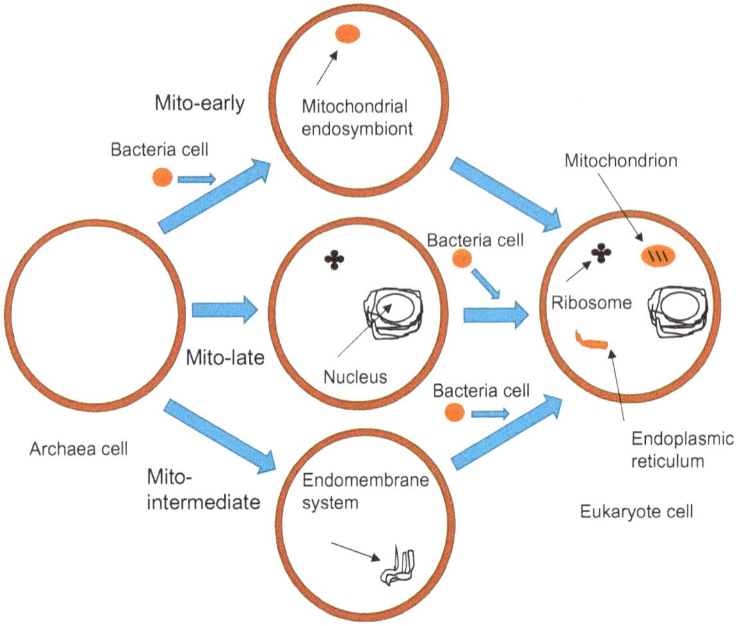

Fig. 8.3.1 Origin of Eukaryote cells and mitochondria (Ettema, 2016). The origin of eukaryote cells are divided into "Mito-early," "Mito-intermediate," and "Mito-late" models.

advantage that greater energy can be produced. One of the questions is when the evolution of the host cell occurred. There are three models: mitochondria is an early event ("mito-early"); the complexity of the eukaryotic cell was already established ("mito-late"); or evolution occurred after the endomembrane system was formed ("mito-intermediate") (see Fig. 8.3.1; Ettema, 2016).

By tracing phylogenetic signals of proteins that were present in the last eukaryotic common ancestor (LECA), Pittis and Gabaldón (2016) identified different classes of protein according to the timing of their appearance in eukaryotes. The most recently acquired LECA proteins are dominated by bacterial proteins, which are primarily located in mitochondria. Pittis and Gabaldón (2016) identified a bacterial LECA protein that was acquired before these mitochondrial proteins. Several of these proteins seem to have existed in intracellular membrane systems, such as the endoplasmic reticulum and the Golgi apparatus. The host cell was already chimaeric before the mitochondrial endosymbiosis. Several bacterial proteins that pre-date the mitochondrial endosymbiosis already worked in intracellular membrane systems. This suggests that the host cell already had a considerable degree of complexity, which is supportive of a late mitochondrial origin ("mito-late") (Pittis and Gabaldón, 2016).

REFERENCES

Diad, M., Bakkeren, E., Cornuault, J.K., Moor, K., Hausmann, A., et al., 2017. Inflammation boosts bacteriophage transfer between *Salmonella* spp. Science 355, 1211–1215.
Ettema, T.J.G., 2016. Mitochondria in the second act. Nature 531, 39–40.
Guo, S., Stevens, C.A., Vance, T.D.R., Olijve, L.L.C., Graham, L.A., et al., 2017. Structure of a 1.5-MDa adhesin that binds its Antarctic bacterium to diatoms and ice. Sci. Adv. 3, e1701440.
Pittis, A.A., Gabaldón, T., 2016. Late acquisition of mitochondria by a host with chimaeric prokaryotic ancestry. Nature 531, 39–40.

CHAPTER NINE

Plant-Related Materials and Topics

Contents

9.1 HOW THE PLANT SOLVES THE DILEMMA THAT POLLINATORS ARE ALSO HERBIVORES

Almost 90% of flowering plants are animal-pollinated (Ollerton et al., 2011). They produce floral scents and other signals to attract pollinators. However, antagonistic visitors, such as herbivores, are also attracted by these floral scents. This dilemma becomes worse when adult insects pollinate the same plant, which is eaten by their larvae. It is unknown how plants maximize their fitness under the contradictory condition.

Zhou et al. (2017) showed that the night-flowering wild tobacco *Nicotiana attenuate* emits a sesquiterpene, (*E*)-α-bergamotene (see Fig. 9.1.1), in flowers, by which adult *Manduca sexta* moth increases pollination. The same compound in leaves is an indirect defense against *M. sexta* larvae (Kessler and Baldwin, 2001; Halitschke et al., 2008).

In more detail, at night, when the nocturnal moth is moving to and fro, the plant's flowers release (*E*)-α-bergamotene. The molecule affects neurons in the insect's proboscis, inducing the moth to stay longer at a flower, which increases the possibility of pollination. During the day, when the adults are resting but the moth's larvae are actively searching for food, the plant leaves produce (*E*)-α-bergamotene. That change attracts predators that feed on the larvae and eggs, resulting in indirect protection of the plant. By optimally coordinating the timing and place of (*E*)-α-bergamotene biosynthesis, the plant has evolved a chemical strategy that increases pollination and defense against herbivores.

Biochemistry for Materials Science
https://doi.org/10.1016/B978-0-12-817054-0.00009-6
© 2019 Elsevier Inc.
All rights reserved.

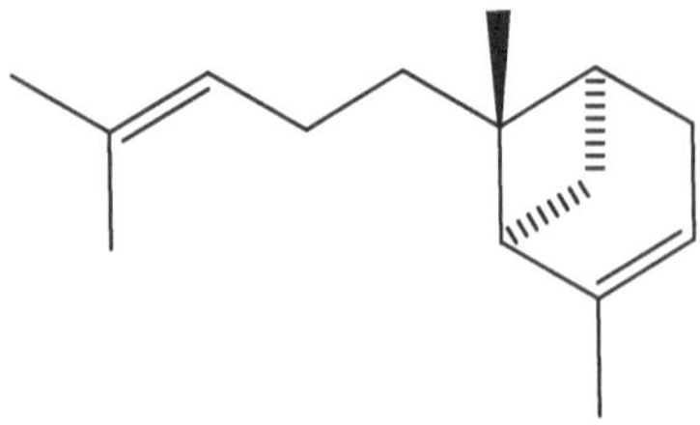

Fig. 9.1.1 (E)-α-bergamotene.

Forward and reverse genetic analyses showed that both herbivory-induced and floral (*E*)-α-bergamotene are controlled by the expression of a monoterpene-synthase-derived sesquiterpene synthase (NaTPS38). The expression pattern of NaTPS38 also explains variation in (*E*)-α-bergamotene emission.

These results highlight that differential expression of a single gene that results in tissue-specific emission of one compound contributes to resolving the dilemma for plants when their pollinators are also herbivores. Furthermore, the study by Zhou et al. (2017) provided genetic evidence that pollinators and herbivores interactively shape the evolution of floral signals and plant defense.

9.2 POSSIBLE PLANT-DERIVED CONTRACEPTIVES

When the sperm get near an egg, their tails start whipping energetically. This final effort improves the sperm's chances of penetrating the viscous surroundings around the egg. This rush is called CatSper, which is based on the calcium channel of sperm. The steroid hormone progesterone activates CatSper of human sperm via binding to the serine hydrolase ABHD2. However, the steroid specificity of ABHD2 has not been evaluated.

Mannowetza et al. (2017) explored which steroid hormones cause CatSper for human sperm. The results show that testosterone, estrogen, and hydrocortisone did not affect CatSper, while the neurosteroid pregnenolone sulfate showed similar effects to progesterone.

Mannowetza et al. (2017) also found that steroid-like molecules such as the plant triterpenoids pristimerin (found in a vine used in traditional Chinese medicine; see Fig. 9.2.1A) and lupeol (found in mangoes; see Fig. 9.2.1B) also reduced CatSper activation by inhibiting the enzyme ABHD2.

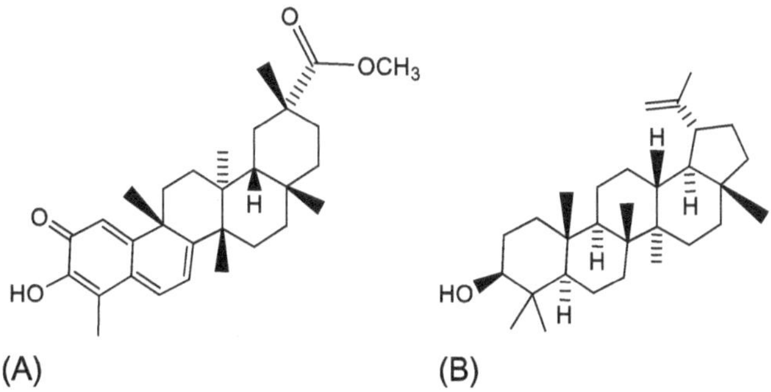

Fig. 9.2.1 Steroid-like triterpenoids from the plant: (A) pristimerin; (B) lupeol.

The pristimerin and lupeol could be contraceptive compounds not only for humans but also for nonhumans.

9.3 COMMUNICATION VIA PARASITIC PLANTS

Dodder (*Cuscuta australis*), also known as wizard's net, devil's guts, witch's hair, etc., is a nefarious rootles and leafless parasitic plant. It is a plant parasite, sucking water and nutrients from its victims using a penetrating organ named a haustorium and connecting the victims through vasculature.

Hettenhausena et al. (2017) found that a herbivore attack on one of the dodder bridge-connected plants induces gene expression and increases the activity of trypsin proteinase inhibitors. This results in higher resistance to insects and other undamaged but dodder-connected plants as well. This dodder-mediated interplant signaling is rapid, far-reaching (>100 cm), and partly requires the plant hormone jasmonic acid. Although dodders negatively influence their host plants, they may also provide ecologically relevant information-based benefits under certain circumstances.

REFERENCES

Halitschke, R., Stenberg, J.A., Kessler, D., Kessler, A., Baldwin, I.T., 2008. Shared signals—"alarm calls" from plants increase apparency to herbivores and their enemies in nature. Ecol. Lett. 11, 24–34.

Hettenhausena, C., Lia, J., Zhuanga, H., Suna, H., Xua, Y., et al., 2017. Stem parasitic plant Cuscuta australis (dodder) transfers herbivory-induced signals among plants. PNAS. 1704536114, E6703–E6709.

Kessler, A., Baldwin, I.T., 2001. Defensive function of herbivore induced plant volatile emissions in nature. Science 291, 2141–2144.

Mannowetza, N., Millera, M.R., Lishkoa, P.V., 2017. Regulation of the sperm calcium channel CatSper by endogenous steroids and plant triterpenoids. PNAS 114, 5743–5748.

Ollerton, J., Winfree, R., Tarrant, S., 2011. How many flowering plants are pollinated by animals? Oikos 120, 321–326.
Zhou, W., Kügler, A., McGale, E., Haverkamp, A., Knaden, M., et al., 2017. Tissue-specific emission of (**E**)-α-bergamotene helps resolve the dilemma when pollinators are also herbivores. Curr. Biol. 27, 1336–1341.

CHAPTER TEN

Animal-Related Materials and Topics

Contents

10.1 GECKOS-INSPIRED ADHESION CONTROLLED MATERIAL

Geckos can walk on vertical or upside-down surfaces. The secret is half a million fibers on their toes, called setae. Each seta has a spatula-like nanostructure, which grips a surface by weak van der Waals forces. This force is collected and strong adhesion can be achieved to support a gecko upside down. Furthermore, fast detachment is also available by the leg movement.

Kizilkan et al. (2017) were inspired by this mechanism and developed an artificial photocontrollable microstructure device. They have already developed a gecko-like adhesion microstructure, which has numerous mushroom-shaped polydimethylsiloxanes. Kizilkan et al. made another layer on the adhesion layer. The layer is made of a stretchy liquid crystalline

Biochemistry for Materials Science
https://doi.org/10.1016/B978-0-12-817054-0.00010-2
© 2019 Elsevier Inc.
All rights reserved.

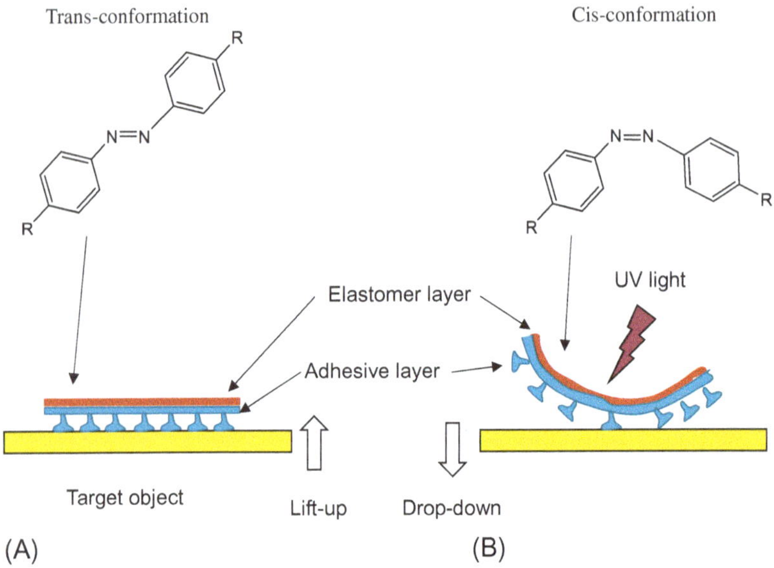

Fig. 10.1.1 Schematic diagrams of a bio-inspired photo-adhesion controlling system. The transportation device is made of two layers: the adhesive layer and the elastomer layer. There are many mushroom-shaped microstructures, which hold the target object using van der Waals forces. The elastomer layer contains azobenzene molecules. (A) The azobenzene molecule is usually trans-conformation. The elastomer layer is flat and the adhesive layer can lift up the target object. (B) When the UV light is turned on, the azobenzene molecule takes cis-conformation, which bends the elastomer layer, and the number of the mushroom structure holding the target object decreases, resulting in dropping down the target object.

elastomer containing azobenzene molecules. Azobenzene transforms from trans to cis under the UV light, which causes the azobenzene layer to bend, resulting in the adhesive layer lifting up. When the UV light is turned off, the layers return to their normal form. The difference of adhesive ability or force is a factor of 2.7 (see Fig. 10.1.1).

This light-controllable device was made for a pick-up and drop-down system for transporting planar and three-dimensional solid objects.

10.2 TERMITE-HUNTING ANTS SEND A CHEMICAL SOS

10.2.1 Introduction

Throughout the animal kingdom from social insects to primates, rescue behavior is observed (Hollis and Nowbahari, 2013). Helping behavior observed in ants can arise in predator-prey interactions—for example, by helping nestmates that have fallen into an ant lion trap by pulling the ant out (e.g., Nowbahari et al., 2009). This rescue behavior in social insects was always directed toward individuals under an urgent threat—that is, suffocation or being eaten.

Megaponera analis is a strictly termitophagous ponerine ant species, in sub-Saharan Africa from 25°S to 12°N (Schmidt and Shattuck, 2014). A scout ant that returns to its nest after finding an active termite foraging site initiates an attack. It will recruit approximately 200–500 nestmates and lead them to the termites in a column-like march formation, which can be up to 50m away from the nest. During the attack, division of labor occurs (Corbara and Dejean, 2000): larger ants (majors) break open the protective soil cover made by the termites, while the smaller ants (minors) go into these openings to kill and pull out the prey (Villet, 1990). Afterward, the majors collect the dead termites, the column forms again, and the hunting party goes back to the nest. These attacks occur two to four times a day.

Termites have various ways to protect themselves effectively against predators such as *M. analis*, of which a specialized soldier caste with strong heads and big mandibles is the main defensive force. Consequently, ants involved in the hunting process have high injury risks.

10.2.2 Helping Behavior in *M. analis*

Frank et al. (2017) observed a unique helping behavior in *M. analis* to compensate for the high injury rate by carrying back injured ants to the nest. The carrying of ants after the hunt was also observed in Kenya and the Democratic Republic of Congo; however, no attempt was made in those studies to explore this behavior. Frank et al. (2017) further observed the removal of termites, still biting on ant extremities, and the rescue behavior of ants carrying long-term injuries of extremity loss. Such special rescue behavior is unexpected in insects, where the value of each individual is generally underestimated.

10.2.3 Pheromone Triggering Rescue Behavior

Initially, stridulation was ruled out in searching for the signal triggering this rescue behavior, although it is known as a mechanism to trigger helping in other ants. Frank et al. (2017) observed that artificially injured ants, on which stridulation was inhibited, were still rescued. After extensive behavioral experiments on dummies (frozen minors), Frank et al. (2017) determined that the mandibular gland should be the most likely candidate. Hindgut content and the poison gland were ruled out as triggers of the behavior.

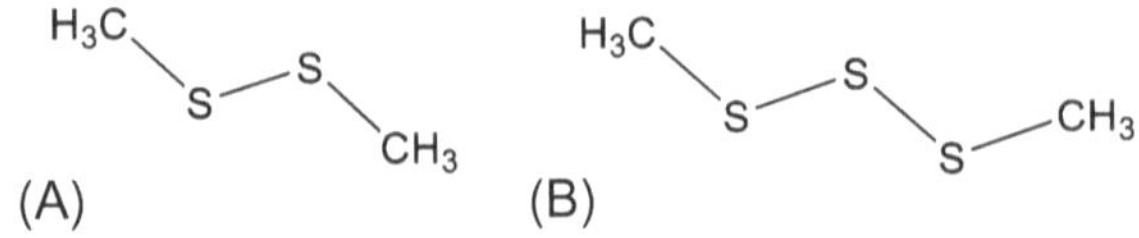

Fig. 10.2.1 The pheromone triggering rescue behavior inside the mandibular gland: (A) dimethyl disulfide (DMDS); (B) dimethyl trisulfide (DMTS).

To test this hypothesis further, when the mandibular gland contents were put on to healthy individuals, healthy ants were carried back by their nestmates. A gas chromatography-mass spectrometry (GC-MS) analysis showed dimethyl disulfide (DMDS; see Fig. 10.2.1) and dimethyl trisulfide (DMTS; see Fig. 10.2.1) to be the main chemical composition of the gland, which were confirmed by a previous analysis of the gland contents and concentrations (14 ng of DMDS and 5 ng of DMTS per gland; Longhurst et al., 1979). Although 9 ng of DMDS alone was not enough to start the rescue behavior on a dummy, 9 ng of DMTS by itself was enough. A clearer response was shown with an equal mixture of the two components (9 ng of DMDS and DMTS each).

After their lives are saved by returning their nest, they heal their wounds, and learn how to walk with fewer extremities. Then many of the wounded ants come back to the attack again.

10.3 BUMBLEBEES CAN DISCRIMINATE SCENT-MARKS BY THE SAME SPECIES

Bumblebees secrete a foot odor, wherever they land, which can be detected by conspecifics. These secretions are called scent-marks, and bumblebees can use them as social signals. Bumblebees can detect and associate scent-marks with rewarding or unrewarding flowers. However, their ability at discriminating between scent-marks from bumblebees of differing relatedness is unknown.

Three experiments were performed by Pearce et al. (2017) for bumblebees (*Bombus terrestris*), where they were repeatedly exposed to rewarding and unrewarding artificial flowers simultaneously. Each flower type carried scent-marks from conspecifics of differing relatedness or was unmarked.

Pearce et al. (2017) found that bumblebees are able to distinguish between:

1. unmarked flowers and scent-marked flowers by themselves;

2. scent-marked flowers by themselves and scent-marked flowers by others in their nest (nestmates); and
3. scent-marked flowers by their nestmates and scent-marked flowers by non-nestmates.

The bumblebees found it more difficult to discriminate between each of the flower types when both flower types were scent-marked. The study of Pearce et al. (2017) showed that bumblebees have the ability to discriminate between scent-marks of the nestmates, which are potentially very similar in their chemical composition, being composed mainly of (*Z*)-9-Tricosene or 9*Z*-Tricos-9-ene (or $C_8H_{17}CH{=}CHC_{13}H_{27}$). This ability prevents nestmates and other bumblebees from visiting the same nectar source.

10.4 HOW NAKED MOLE-RATS SURVIVE UNDER ANOXIA

African naked mole-rats (*Heterocephalus glaber*) are social animals living underground. They live in a colony (see Fig. 10.4.1), which may be composed of up to 300 members. Therefore, they are always in danger of lack of oxygen. In experimental conditions, they tolerate hours under extreme low oxygen, and survive 18 min under no oxygen (anoxia) conditions.

During anoxia, the naked mole-rat changes its metabolism from glycolysis using glucose (see Fig. 10.4.2B) to anaerobic metabolism fueled by fructose (see Fig. 10.4.2A), which is actively gathered and changed into lactate in the brain. For other mammals, low oxygen conditions means death, because glycolysis, which produces adenosine triphosphate (ATP) for energy

Fig. 10.4.1 A photograph of a colony of naked mole-rats (*Heterocephalus glaber*). *Copyright: Wikipedia, Edward Russell at flickr: https://upload.wikimedia.org/wikipedia/commons/1/1f/Naked_Mole_Rats-cropped.jpg.*

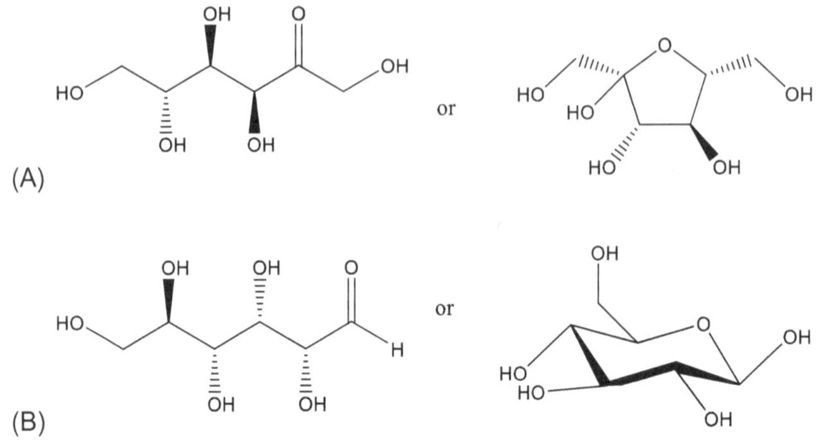

Fig. 10.4.2 (A) Fructose. (B) Glucose.

production, is stopped at once. No ATP means cells are out of energy and die, resulting in the death of an individual.

Park et al. (2017) observed the brain tissue of the naked mole-rat after 5% O_2 conditions injecting ^{13}C-fructose. The glycolysis and fructolysis pathways of the naked mole-rat are shown in Fig. 10.4.3 as GLUT1 and GLUT5, respectively. Glucose enters the brain cell via GLUT1. Glucose is finally converted via phosphofructokinase (PFK), which is a rate-limiting step. Fructose enters the brain cell via GLUT5 and is phosphorylated by ketohexokinase (KHK) to fructose-1-phosphate (Fructose-1-P) at a much higher efficiency than by hexokinase (HK). Fructose-1-P is directly metabolized into trioses via aldolase B (ALDOB) or aldolase C (ALDOC), bypassing feedback inhibition. DHAP and Glyc-3-P are dihydroxyacetone phosphate and glycerol-3-phosphate, respectively. Glyceraldehyde-3-phosphate (GA3P) and tricarboxylic acid (TCA) are observed as the intermediates. However, glycolytic end-products like pyruvate and lactate are also observed. Global expression of the GLUT5 fructose transporter and high levels of ketohexokinase (KHK) were identified as molecular signatures of the fructose metabolism.

In summary, the naked mole-rat has evolved the ability to use fructose to fuel the brain under low oxygen conditions. This metabolism contains transporters and enzymes that metabolize fructose to lactate using a pathway that bypasses using PFK (Fig. 10.4.3, GLUT1). Fructose and sucrose are both increased in the naked mole-rat during anoxia. The source of these sugars

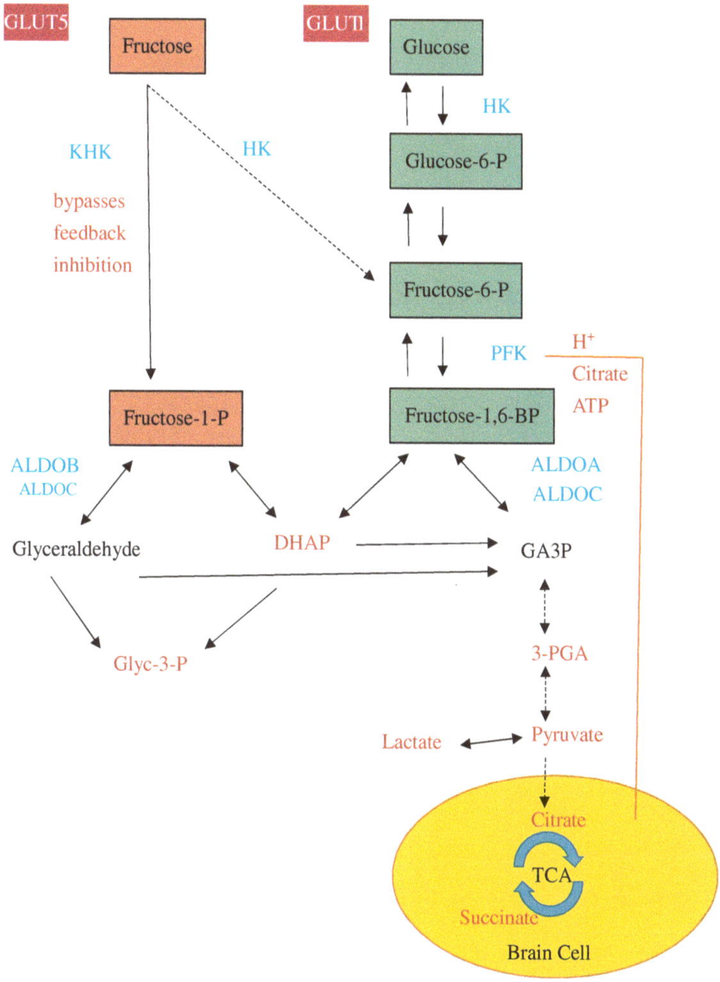

Fig. 10.4.3 Metabolic flow of fructose metabolites in the hypoxic brain. GLUT5 and GLUT1 are fructolysis and glucolysis pathways, respectively. See the text for details.

is unknown. Fructolysis in mammals is usually restricted to the liver and kidney (Port et al., 2012). A change to fructolysis under low oxygen conditions is related to such health issues as cancer malignancy or metabolic syndrome. It is thus important to understand how naked mole-rats use fructolysis with no apparent physiological disadvantage. Molecular studies of the naked mole-rat may reduce hypoxic damage caused by ischemic heart disease and stroke.

10.5 ANOTHER CREATURE EATS PLASTIC

Polyethylene (PE) represents ~40% of total demand for plastic products. PE is discarded in landfills or used in energy recovery by combustion, and only ~30% is used in recycling. Furthermore, PE shopping bags are a source of environmental pollution, because they are used widely due to their low price and convenience. Therefore, new methods for plastic degradation, especially of PE, are urgently required.

Bombelli et al. (2017) observed fast bio-degradation of PE into ethylene glycol by larvae of the wax moth *Galleria mellonella*. When a PE film is left in direct contact with the wax worms, holes start to appear after 40 min., and a mass loss becomes ~92 mg/12 h/100 worms. When the treated PE is eaten, the degradation rate becomes 0.23 mg cm^{-2} h^{-1}. This is a far higher rate than that of PET microbial degradation (Yoshida et al., 2016).

The point at which the worm smeared was analyzed by FTIR, and the peak of ethylene glycol as well as the carbonyl bond was detected. The wax worm feeds on beeswax, which is made of a highly diverse mixture of lipid compounds, such as alkanes, alkenes, fatty acids, and esters (Maia and Nunes, 2013). The most frequent hydrocarbon bond is CH_2-CH_2, as in PE.

Further investigation is urgent to analyze the nature products and also gene coding for enzymes.

10.6 ARTIFICIAL SPIDER SILK PRODUCTION

Spider silk is a very special material due to its strength. However, spiders do not produce enough fibers to be used commercially. Therefore, many researchers are trying to produce them by bioengineering techniques in modified cell or animals. In these methods, the fibers are made as water-insoluble proteins, and then spun into fibers in severe conditions, which is far from the process that spiders use.

Andersson et al. (2017) have produced spider silk by a much closer process to the one that spiders use. Spider silk proteins are made of an N-terminal domain and a C-terminal domain bracketing a region of repeated amino acids, which make the silk strong (Hagn et al., 2010). The N-terminal domains are very water-soluble, while some C-terminal domains are not soluble at all.

Andersson et al. (2017) used N-terminal and repetitive regions from one spider species, and relatively soluble C-terminal regions from another species. They produced proteins from bacteria with solubility of 500 mg/mL. In order to "spin" nearly a kilometer of silk fiber, they pumped a pH 7.5 solution of the protein (spidroin) into a pH 5 aqueous buffer, mimicking the process that spiders use.

10.7 WHY ARE POISON FROGS RESISTANT TO BATRACHOTOXIN?

Poison-dart *Phyllobates terribilis* frogs accumulate lethal amounts of steroidal alkaloid batrachotoxin (BTX; see Fig. 10.7.1) separately in their skin as a defense mechanism. The average level of BTX in one frog is 1 mg, which could kill 25,000 mice. BTX targets voltage-gated Na^+ channels and opens permanently. Why such frogs are resistant to BTX remained a mystery for some time.

Wang and Wang (2017) solved this mystery. They found that just one amino acid swap (asparagine → threonine), which can be caused via a single nucleotide mutation (AAC → ACC; see Table 2.8.1) in the *P. terribilis* muscle of Na^+ channel, can give BTX resistance.

10.8 A SINGLE AMINO ACID VARIATION MAKES DIFFERENCES IN CETACEAN SIZES

Cetaceans, including whales, dolphins, and porpoises, a group of mammals adapted to the aquatic environment that descended from terrestrial

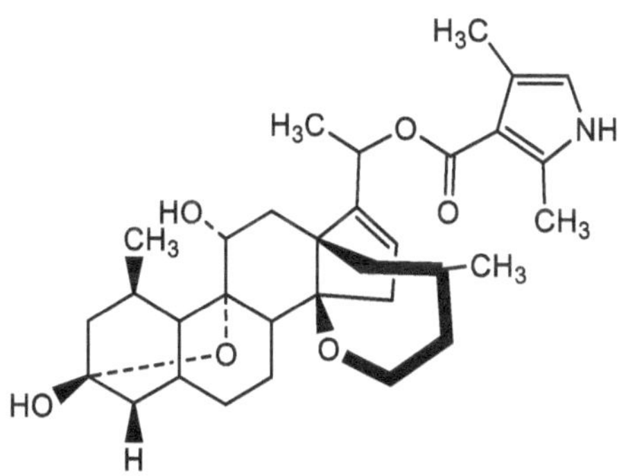

Fig. 10.7.1 Batrachotoxin (BTX).

artiodactyls, exhibit tremendous interspecific differences in a number of phenotypes, including feeding behavior, such as filter feeding in the Mysticeti vs prey-hunting Odontoceti, and size, with the smallest cetacean, the vaquita, at 1.4 m. and the largest, the blue whale, reaching 33 m.

The Melanocortin-4 receptor (MC4R) regulates food intake, energy balance, and somatic growth in both mammals and teleosts. Zhao et al. (2017) examined allelic variants of the MC4R in cetaceans. They sequenced the MC4R from 20 cetaceans, and pharmacologically characterized 17 of these protein products.

Results identified a single variation at amino acid 156 in the MC4R. The representative species of major cetacean lineages are uniquely associated with the toothed whales or Odontoceti (arginine at 156; R156) and baleen whales or Mysticeti (glutamine at 156; Q156).

The Q156 receptor variant found in the larger baleen whales was functionally less responsive to its endogenous anorexigenic ligand, α-MSH. In addition, the R156 receptor variant showed greater constitutive activity and a higher affinity for ligand. These data suggest that the MC4R may be one gene involved in the evolution of feeding ecology, energy balance, and body size in cetaceans.

10.9 TOUGH ADHESIVES (TAs)

Adhesion to wet and dynamic surfaces, such as biological tissues, is important in many fields; however, it is extremely challenging. Existing adhesives are cytotoxic, stick weakly to tissues, or cannot be used in wet environments.

Li et al. (2017) developed tough adhesives (TAs) made of two layers. The design was inspired by a mucus secreted by slugs (*Arion subfuscus*), which strongly adheres to wet surfaces (Pawlicki et al., 2004). This slug adhesive consisted of a tough matrix with interpenetrating positively charged proteins (Wilks et al., 2015). TA developed by Li et al. (2017) is composed of an adhesive surface (chitosan + EDC*/sulfo-NHS**) and a dissipative matrix of alginate-polyacrylamide (alginate-PAAm) (see Fig. 10.9.1A). The former adheres to the substrate by electrostatic interactions, covalent bonds, and physical interpenetration by chitosan + EDC/sulfo-NHS (see Fig. 10.9.1A–D). The latter amplifies energy dissipation through hysteresis under deformation (see Fig. 10.9.1A, E, and F). The two layers

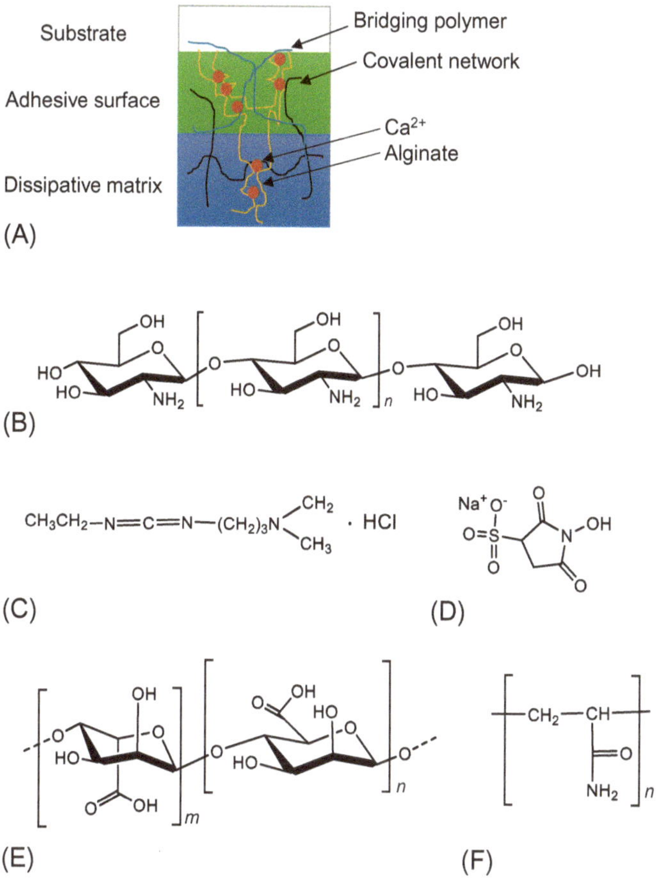

Fig. 10.9.1 Tough adhesives (TAs): (A) a schematic figure of TAs on substrate; (B) chitosan; (C) 1-ethyl-3-(3-dimethylaminopropyl) carbodiimide hydrochloride (EDC); (D) N-hydroxysulfosuccinimide sadium salt (sulfo-NHS); (E) alginate; (F) polyacrylamide.

synergistically lead to higher adhesion energies on wet surfaces, compared with those of existing adhesives. Adhesion occurs within minutes, independent of blood exposure and compatible with in vivo dynamic movements.

This family of adhesives may be useful in many areas of application, including tissue adhesives, wound dressings, and tissue repair.

(*) EDC: 1-Ethyl-3-(3-dimethylaminopropyl) carbodiimide hydrochloride (see Fig. 10.9.1C).

(**) sulfo-NHS: N-hydroxysulfosuccinimide sodium salt (see Fig. 10.9.1D).

10.10 WHY FENETHYLLINE IS ABUSED BY YOUNG PEOPLE

Fenethylline (see Fig. 10.10.1A), also known by the name Captagon, is abused by young people, terrorists and criminal organizations to enhance fearlessness, aggression, etc.

Wenthur et al. (2017) developed a "dissection through vaccination" approach, called DISSECTIV. They added fenethylline to mice. Then they added each vaccine, which conjugates to fenethylline, theophylline (Fig. 10.10.1B), and amphetamine (Fig. 10.10.1C), and stops each psychic effect. Theophylline and amphetamine are the possible metabolite compounds of fenthylline.

Wenthur et al. found theophylline and amphetamine activities in mice. Theophylline blocks adenosine receptors in the brain, and amphetamine enhances dopamine signaling. By taking fenethylline, theophylline and amphetamine effects appeared. This is likely to be why fenethylline is abused by young people or terrorists.

Thus DISSECTIV is very effective to expose unidentified active chemical species and resolve pharmacodynamic interactions within other chemically complex systems.

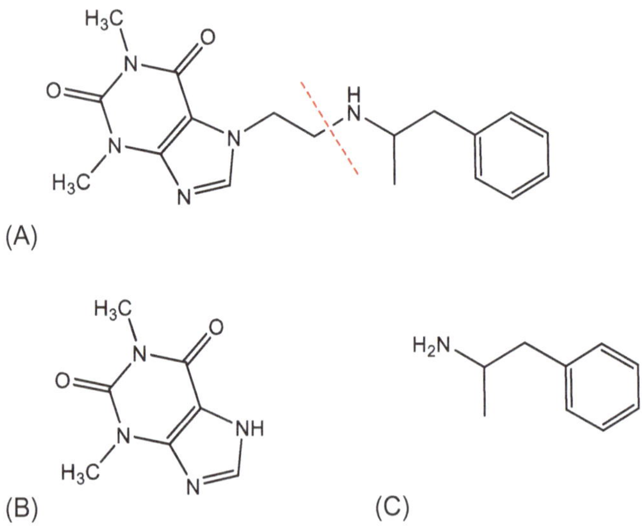

Fig. 10.10.1 (A) Fenethylline. The dashed line shows where an enzyme splits fenethylline into (B) and (C). (B) Theophylline. (C) Amphetamine.

REFERENCES

Andersson, M., Jia, Q., Abella, A., Lee, X.-Y., Landreh, M., et al., 2017. Biomimetic spinning of artificial spider silk from a chimeric minispidroin. Nat. Chem. Biol. 13, 262–264.

Bombelli, P., Howe, C.J., Bertocchini, F., 2017. Polyethylene bio-degradation by caterpillars of the wax moth *Galleria mellonella*. Curr. Biol. 27, R292–R293.

Corbara, B., Dejean, A., 2000. Adaptive behavioral flexibility of the ant Pachycondyla analis (=Megaponera foetens) (Formicidae: Ponerinae) during prey capture. Sociobiology 36, 465–483.

Frank, E.T., Schmitt, T., Hovestadt, T., Mitesser, O., Stiegler, J., Linsenmair, K.E., 2017. Sci. Adv. 3, e1602187.

Hagn, F., Eisoldt, L., Hardy, J.G., Vendrely, C., Coles, M., et al., 2010. A conserved spider silk domain acts as a molecular switch that controls fibre assembly. Nature 465, 239–242.

Hollis, K.L., Nowbahari, E., 2013. Toward a behavioral ecology of rescue behavior. Evol. Psychol. 11, 647–664.

Kizilkan, E., Strueben, J., Staubitz, A., Gorb, S.N., 2017. Bioinspired photocontrollable microstructured transport device. Sci. Robot. 2. https://doi.org/10.1126/scirobotics.aak9454.

Li, J., Celiz, A.D., Yang, J., Yang, Q., Wamala, I., et al., 2017. Tough adhesives for diverse wet surfaces. Science 357, 378–381.

Longhurst, C., Baker, R., Howse, P.E., 1979. Termite predation by *Megaponera foetens* (FAB.) (Hymenoptera: Formicidae). J. Chem. Ecol. 5, 703–719.

Maia, M., Nunes, F.M., 2013. Authentication of beeswax (Apis mellifera) by high-temperature gas chromatography and chemometric analysis. Food Chem. 136, 961–968.

Nowbahari, E., Scohier, A., Durand, J.-L., Hollis, K.L., 2009. Ants, cataglyphis cursor, use precisely directed rescue behavior to free entrapped relatives. PLoS ONE 4, e6573. https://doi.org/10.1371/journal.pone.0006573.

Park, T.J., Reznick, J., Peterson, B.L., Blass, G., Omerbašić, D., et al., 2017. Fructose-driven glycolysis supports anoxia resistance in the naked mole-rat. Science 356, 307–311.

Pawlicki, J.M., Pease, L.B., Pierce, C.M., Startz, T.P., Zhang, Y., et al., 2004. The effect of molluscan glue proteins on gel mechanics. J. Exp. Biol. 207, 1127–1135.

Pearce, R.F., Giuggioli, L., Rands, S.A., 2017. Bumblebees can discriminate between scent-marks deposited by conspecifics. Sci. Rep. 7, 43872.

Port, A.M., Ruth, M.R., Istfan, N.W., 2012. Fructose consumption and cancer: is there a connection? Curr. Opin. Endocrinol. Diabetes Obes. 19, 367–374.

Schmidt, C.A., Shattuck, S.O., 2014. The higher classification of the ant subfamily Ponerinae (Hymenoptera: Formicidae), with a review of Ponerine ecology and behavior. Zootaxa 3817, 1–242.

Villet, M.H., 1990. Division of labour in the Matabele ant Megaponera foetens (Fabr.) (Hymenoptera Formicidae). Ethol. Ecol. Evol. 2, 397–417.

Wang, S.-Y., Wang, G.K., 2017. Single rat muscle Na^+ channel mutation confers batrachotoxin autoresistance found in poison-dart frog *Phyllobates terribilis*. PNAS. pnas.1707873114.

Wenthur, C.J., Zhou, B., Janda, K.D., 2017. Vaccine-driven pharmacodynamic dissection and mitigation of fenethylline psychoactivity. Nature 548, 476–479.

Wilks, A.M., Rabice, S.R., Garbacz, H.S., Harro, C.C., Smith, A.M., 2015. Double-network gels and the toughness of terrestrial slug glue. J. Exp. Biol. 218, 3128–3137.

Yoshida, S., Hiraga, K., Takehana, T., Taniguchi, I., Yamaji, H., et al., 2016. A bacterium that degrades and assimilates poly(ethylene terephthalate). Science 351, 1196–1199.

Zhao, L., Zhou, X., Rokas, A., Cone, R.D., 2017. Functional variants of the melanocortin-4 receptor associated with the Odontoceti and Mysticeti suborders of cetaceans. Sci. Rep. 7, 5684. https://doi.org/10.1038/s41598-017-05962-1.

FURTHER READING

Hollis, K.L., Harrsch, F.A., Nowbahari, E., 2015. Ants vs. antlions: an insect model for studying the role of learned and hard-wired behavior in coevolution. Learn. Motiv. 50, 68–82.

CHAPTER ELEVEN

Protein and Enzyme-Related Topics

Contents

11.1 AMYLOID STRUCTURE FORMED BY α-HELIX

Amyloid fibrils are protein aggregates that form long strands. Many of these amyloids fibrils are related to human diseases. For example, the amyloid-β plaques and the α-synuclein aggregates are involved in Alzheimer's and Parkinson's diseases, respectively.

The β-sheet structures shown in Fig. 11.1.1A are so popular and ubiquitous among amyloids that many researchers consider that formation of β-fibrils is required for the diseases to occur. The bacterium *Staphylococcus aureus* produces PSMα3, which is the most cytotoxic member of peptides. For the producing bacterium, PSMα3 is not harmful but beneficial as a functional amyloid. For human cells, however, PSMα3 is the cause of virulence of *Staphylococcus aureus*. During the determination of crystal structure of PSMα3, Tayeb-Fligelman et al. (2017) found a new amyloid like α-fibril structure in PSMα3, which was made by α-helix (see Fig. 11.1.1B). In both β-fibril and α-fibril amyloid-like structures, the peptides align perpendicular to the fibril axis (see Fig. 11.1.1).

This bacterium is famous for methicillin-resistant *S. aureus* (MRSA). This resistance should be related to biofilm stability of the α-fibrils produced

Biochemistry for Materials Science
https://doi.org/10.1016/B978-0-12-817054-0.00011-4

© 2019 Elsevier Inc.
All rights reserved.

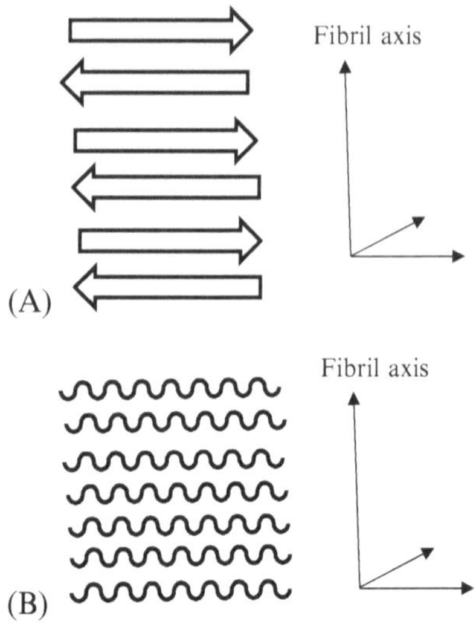

Fig. 11.1.1 Two types of amyloid fibrils. (A) A schematic figure of β-sheet structure found in yeast prion Sup35 (PDB 1YJO). (B) A schematic figure of α-helical sheet found in PSMα3. The wavy symbol indicates α-helix.

by *S. aureus*; however, future studies are required to reveal the relation between the function of the α-fibrils and their stability. It is also unclear whether this is a unique example of the α-fibrils in nature or not.

11.2 INDOLE ALKALOID BIOSYNTHESIS

Stigonematalean cyanobacteria produces hapalindole-type alkaloids, such as hapalindoles and fischerindoles, which are members of the large family of indole monoterpenoids. These have promising antimicrobial and cytotoxic properties, but their syntheses are very difficult. Showing profound biological activities as agents, these natural products exhibit remarkable structural diversities within a given producer or across different producers.

Recently, several biosynthetic gene clusters (BGCs) for hapalindole-type alkaloids were discovered, and the molecular mechanisms were clarified underlying the origin and evolution of the structural diversities of these complex indole alkaloids.

Zhu and Liu (2017) discovered that U-protein-enzymes catalyze cascade reactions. The enzymes synthesize a common precursor into different indole alkaloids. For example, a single indolenine (3-geranyl 3-isocyanovinyl indolenine) is synthesized into a tetracyclic fisherindole (12-*epi*-fischerindole U), a tricyclic hapalindole (12-*epi*-hapalindole C) and a tetracyclic hapalindole (12-*epi*-hapalindole U) by three types of enzymic cascades utilizing WelU1, WelU3, and AmbU4, respectively (see Fig. 11.2.1). Each enzyme's cascade reaction involves three steps: a Cope rearrangement, an aza-Prins cyclization and final carbocation deposition stages. There are three types of steps in the final stages, which make three different indole alkaloids described above. The indolenine precursor and its derivatives are structural isomers, so that the cascades are multistep rearrangements. With U-protein enzymes now in hand – as they can be isolated from cyanobacteria—Liu believes it will be easier to synthesize diverse members of this family of bioactive natural products in the laboratory.

This discovery by Zhu and Liu (2017) gives strong evidence that the U-family proteins encoded in all hapalindole-type alkaloid BGCs are responsible for the first layer of structural diversification of this family of

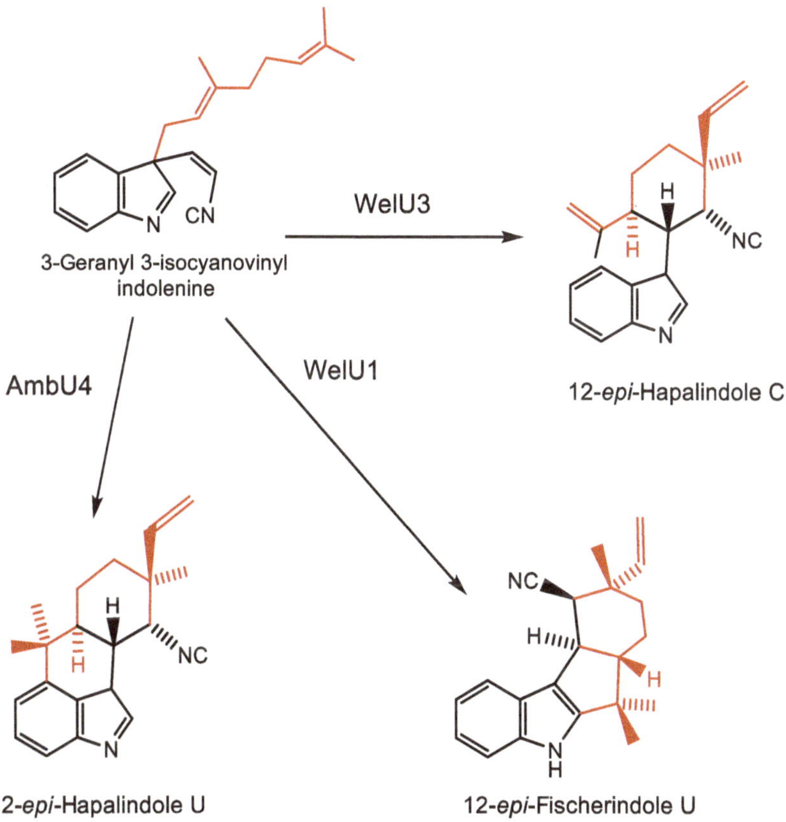

Fig. 11.2.1 The single indolenine precursor is synthesized into different indoles alkaloids by different enzymes.

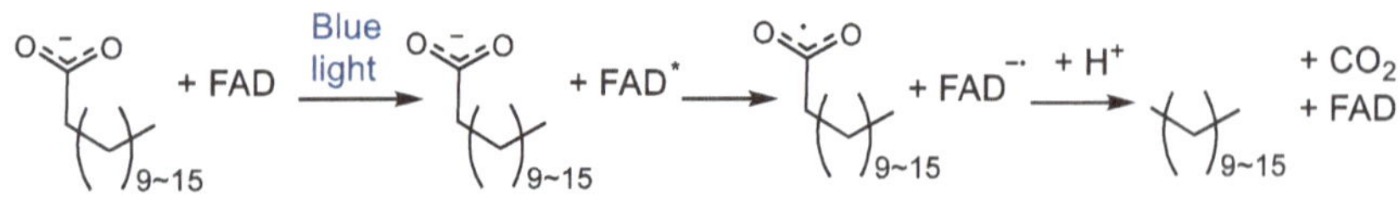

Fig. 11.3.1 Lipid metabolism by newly found photoenzyme.

alkaloids. The ability to reconstitute the U-family enzymes in vitro, which can be easily purified from *Escherichia coli*, will facilitate further mechanistic and structural studies of this family of novel enzymes as well as bioactive natural products.

11.3 NEW PHOTOENZYME CONVERTS FATTY ACIDS TO HYDROCARBONS

Many organisms capture or react to sunlight; however, few enzymes are known to work by light. Among these are DNA photolyases and the photosynthetic reaction centers.

Sorigué et al. (2017) have found that the microalga *Chlorella variabilis* NC64A hides a photoenzyme that acts in lipid metabolism. This enzyme belongs to an algae-specific clade of the glucose-methanol-choline oxidoreductase family and catalyzes the decarboxylation of free fatty acids to n-alkanes or -alkenes in response to blue light (see Fig. 11.3.1).

Crystal structure of the protein reveals a fatty acid-binding site in a hydrophobic tunnel leading to the light-capturing flavin adenine dinucleotide (FAD) cofactor. The decarboxylation starts from electron abstraction from the fatty acid by the photoexcited FAD with a quantum yield >80%. This photoenzyme, which is named fatty acid photodecarboxylase, may be useful in light-driven, bio-based production of hydrocarbons.

11.4 ENZYME SYNTHESIS OF PSILOCYBIN

Psilocybin (see Fig. 11.4.1A) is a major chemical constituent of *Psilocybe cyanescens* ("magic mushrooms"), which causes euphoria and hallucinations.

Fricke et al. (2017) sequenced the genomes of *P. cyanescens* and *P. cubensis*, and the latter was chosen as the model. They characterized three psilocybin biosynthesis enzymes: (i) a fungal L-tryptophan decarboxylase, Posed

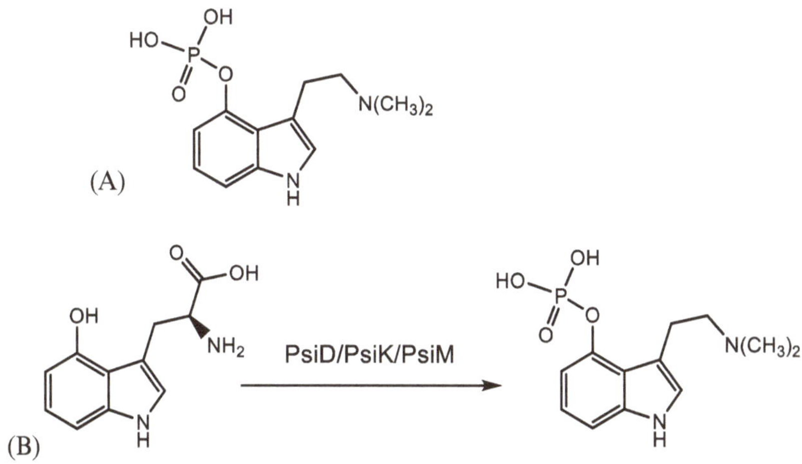

Fig. 11.4.1 (A) Psilocybin. (B) Psilocybin synthetic route from 4-hydroxy-L-tryptophan by Fricke et al. (2017).

(49.6 kDa, 439 amino acids); (ii) a phosphotransferase, PsiK (40.4 kDa, 362 amino acids); and (iii) the methyltransferase, PsiM (34.5 kDa, 309 amino acids). In a combined PsiD/PsiK/PsiM reaction in one pot, psilocybin was synthesized enzymatically from 4-hydroxy-L-tryptophan (see Fig. 11.4.1B).

If pharmaceutical interests in psilocybin exist, their results may give the foundation for its biotechnological production.

11.5 RETICULOCYTE-ERYTHROCYTE TRANSITION

The reticulocyte-red blood cell (erythrocyte) transition (see Fig. 11.5.1) is an example of terminal differentiation. The mature erythrocyte has only hemoglobin concentrated to ~98% of soluble protein. During reticulocyte maturation, the proteome is remodeled through the programmed elimination of most generic constituents of the cell, in parallel with abundant synthesis of hemoglobin. The mechanisms in terminally differentiating cells remain unclear.

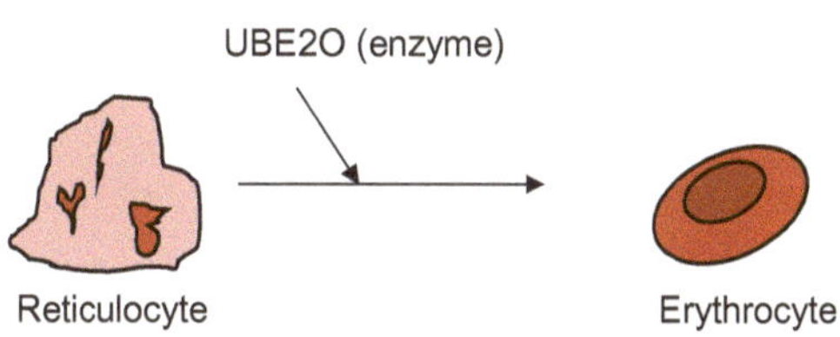

Fig. 11.5.1 UBE2O enzyme drives reticulocyte-erythrocyte transition.

Nguyen et al. (2017) found that UBE2O is a broad-spectrum ubiquitinating enzyme that remodels the erythroid proteome. UBE2O recognized ribosomal proteins and other substrates directly, and targeted them to degradation. Thus, in reticulocytes, the induction of ubiquitinating factors may drive the transition from a complex to a simple proteome.

11.6 CORRECT STRUCTURES OF BAULAMYCINS A AND B

We sometimes happen to meet biologically active molecules in very little quantities that we can never discover again. In such cases, determining the structure of a molecule is paramount, enabling us to understand its biological activity. Molecules that comprise stereochemically complex acyclic and conformationally flexible carbon chains make such a task extremely challenging.

The baulamycins (A and B) are such examples. They were isolated in small quantities and shown to have promising antimicrobial activity. The structures were determined largely through *J*-based configurational analysis (a method based on carbon-proton spin-coupling constants ($^{2,3}J_{C,H}$) and interproton spin-coupling constants ($^{3}J_{H,H}$)) (Matsumori et al., 1999; Tripathi et al., 2014), but this has since been found to be incorrect.

Wu et al. (2017) identified the true structures of the baulamycins by a method for the rapid structural elucidation of such molecules (see Figs. 11.6.1A and B). Especially, they predicted nuclear magnetic resonance (NMR) parameters through density functional theory, and could synthesize the correct structures. In that method, the NMR was combined with an

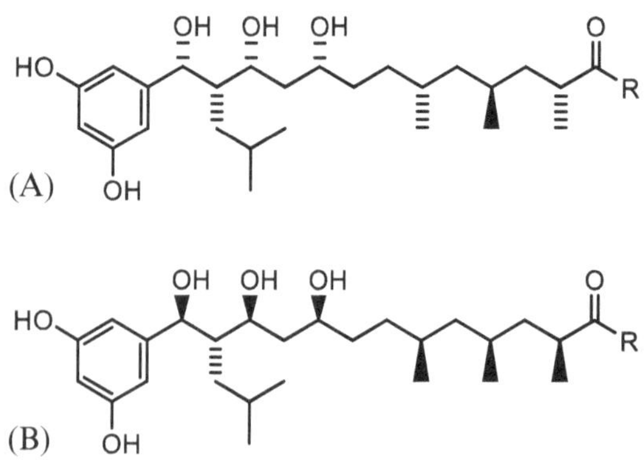

Fig. 11.6.1 Baulamycins A and B (Baulamycin A: R=CH_2CH_3; Baulamycin B: R=CH_3). (A) Previous structures. (B) Revised structures.

efficient sequence of boron-based synthetic transformations, which allowed an encoded (labeled) mixtures of natural-product diastereomers to be prepared.

REFERENCES

Fricke, J., Blei, F., Hoffmeister, D., 2017. Enzymatic synthesis of psilocybin. Angew. Chem. Int. Ed. 56, 12352–12355.

Matsumori, N., Kaneno, D., Murata, M., Nakamura, H., Tachibana, K., 1999. Stereo chemical determination of acyclic structures based on carbon – proton spin-coupling constants. A method of configuration analysis for natural products. J. Org. Chem. 64, 866–876.

Nguyen, A.T., Prado, M.A., Schmidt, P.J., Sendamarai, A.K., Wilson-Grady, J.T., et al., 2017. UBE2O remodels the proteome during terminal erythroid differentiation. Science 357, 471.

Sorigué, D., Légeret, B., Cuiné, S., Blangy, S., Moulin, S., et al., 2017. An algal photoenzyme converts fatty acids to hydrocarbons. Science 357, 903–907.

Tayeb-Fligelman, E., Tabachnikov, O., Moshe, A., Goldshmidt-Tran, O., Sawaya, M.R., et al., 2017. The cytotoxic *Staphylococcus aureus* PSMα3 reveals a cross-α amyloid-like fibril. Science 355, 831–833.

Tripathi, A., Schofield, M.M., Chlipala, G.E., Schultz, P.J., Yim, I., et al., 2014. Baulamycins A and B, broad-spectrum antibiotics identified as inhibitors of siderophore biosynthesis in *Staphylococcus aureus* and *Bacillus anthracis*. J. Am. Chem. Soc. 136, 1579–1586. **erratum** 2014. 136, 10541–10541.

Wu, J., Lorenzo, P., Zhong, S., Ali, M., Butts, C.P., et al., 2017. Synergy of synthesis, computation and NMR reveals correct baulamycin structures. Nature 547, 436–440.

Zhu, Q., Liu, X., 2017. Molecular and genetic basis for early stage structural diversifications in hapalindole-type alkaloid biogenesis. Chem. Commun. 53, 2826–2829.

FURTHER READING

Bhat, V., Dave, A., MacKay, J.A., Rawal, V.H., 2014. The chemistry of hapalindoles, fischerindoles, ambiguines, and welwitindolinones. Alkaloids Chem. Biol. 73, 65–160. 2014.

PART FOUR

Life-Sciences Related Materials and Topics

CHAPTER TWELVE

Pain-Related Materials and Topics

Contents

12.1 PATH OF PAIN

Fig. 12.1.1 schematically depicts the usual path of pain. There are sensors all over the skin which detect pain, heat (hot or cold) or pressure. Each sense is detected by different types of sensory receptor cells. The pain signal is generated by one of these cells, which is called a nociceptor.

The cell of the nociceptor has a very long neuron, and it goes to the spinal cord covered by the backbone. The pain signal is transmitted from the nociceptor cell to the spinal cord at the spinal cell synapse. The spinal cell synapses at the dorsal horn are very complicated, and are controlled by both the sympathoneural and sympathoadrenal signals from the brain. Many painkillers including opioids works on the spinal cell synapses. For example,

Biochemistry for Materials Science
https://doi.org/10.1016/B978-0-12-817054-0.00012-6
© 2019 Elsevier Inc.
All rights reserved.

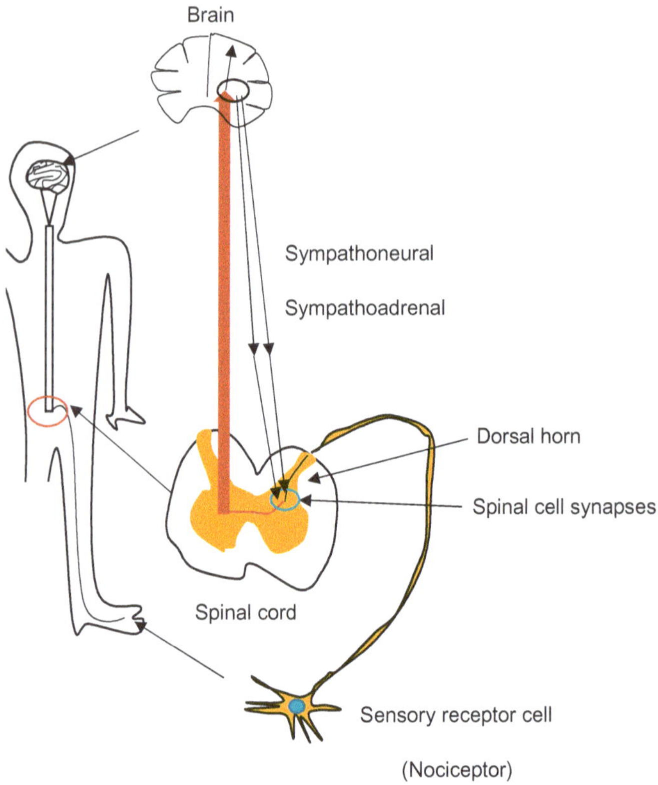

Fig. 12.1.1 Path of pain.

morphine (see Fig. 12.1.2A) works on "μ-opioid receptors" (MORs) at the spinal cell synapses. There are also δ or κ and subtype opioid receptors. Finally, the pain signal goes up to the brain through the spinal cord.

12.2 BLOCKING DOSE-ESCALATION OF OPIOIDS

12.2.1 Prevention of Dose-Escalation

One of the disadvantages of opioids is that the dose required becomes higher and higher; this is called dose-escalation. Corder et al. (2017) reported that the activity of a certain type of sensory neuron drives this phenomenon in mice. The study proposes a new mechanism for two opioid side effects that are responsible for dose escalation: (i) tolerance, in which a patient becomes less sensitive to a given dose of a drug; and (ii) hyperalgesia, in which opioid use paradoxically makes a person more sensitive to pain.

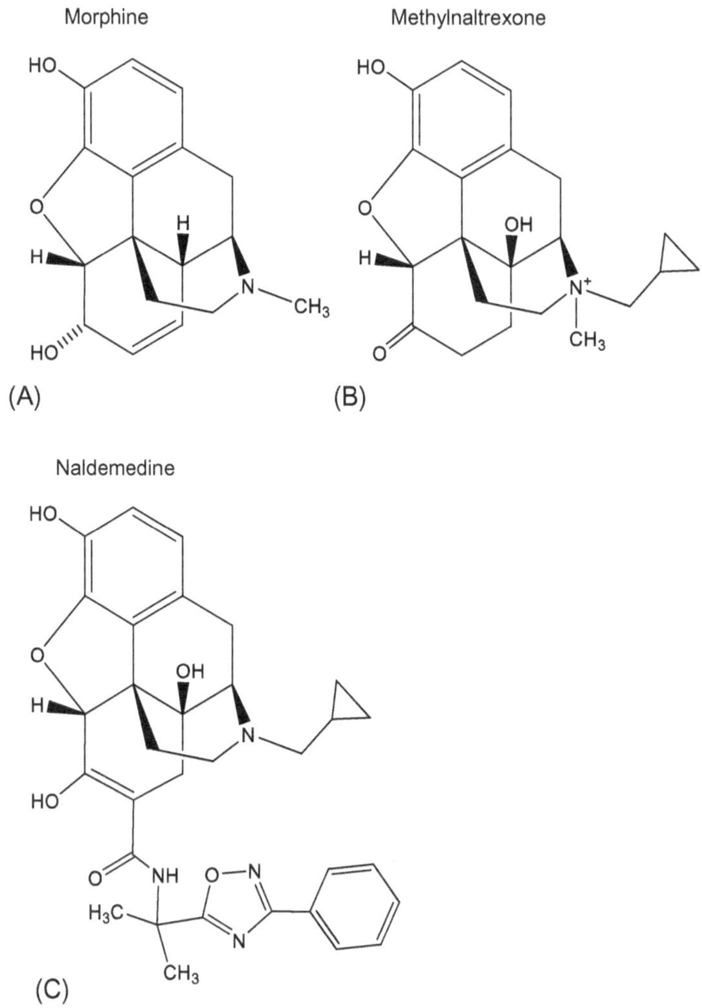

Fig. 12.1.2 (A) Morphine. (B) Methylnaltrexone. (C) Naldemedine.

The mechanism outlined by the team focuses on cells called nociceptors, which relay pain signals from the skin to the spinal cord. They hypothesized that when drugs activate μ-opioid receptors (MORs) on the nociceptors, these trigger signaling pathways that lead to tolerance and hyperalgesia.

To test this hypothesis, the scientists engineered mice of which the mice no longer expressed MORs on their nociceptors. The scientists then observed how effectively a daily dose of the opioid morphine relieved different forms of pain in the mice over 10 days. Nonengineered mice barely responded to the dose by the tenth day by the dose-escalation. However, the engineered mice continued to experience pain relief. The scientists concluded that this demonstrated that MORs in these pain-sensing neurons are necessary for tolerance to develop.

The scientists could prevent tolerance in nonengineered mice with the help of methylnaltrexone (see Fig. 12.1.2B), an FDA-approved drug that blocks MORs (Thomas et al., 2008). Methylnaltrexone cannot cross the blood-brain barrier because of its positively charged quaternary ammonium group. This means that the methylnaltrexone blocks MORs only on cells outside of the brain and spinal cord, such as nociceptors, and not on those within the brain, where opioids trigger most of their pain-relieving effects. Mice given morphine with methylnaltrexone experienced pain relief without developing a tolerance.

Methylnaltrexone is already used for prevention of constipation. As morphine causes constipation, taking methylnaltrexone with morphine has two effects: prevention of constipation as well as dose-escalation.

12.2.2 For Future Application

In Japan, methylnaltrexone is not still approved. However, a similar medicine, naldemedine (see Fig. 12.1.2C), has already been approved there for constipation. Many opioid users who need strong painkillers in Japan hope that either naldemedine will work against the dose-escalation, or that methylnaltrexone will soon be approved and available in Japan. It is known that naldemedine also does not pass through the brain-blood barrier, so naldemedine does not cause the hyperalgesia.

The author has experienced constipation caused by morphine; he took naldemedine hoping that it would also have the side effect of the dose-escalation relief. It worked on constipation; however, the author felt some strange feelings, and thus stopped taking naldemedine after several days. Unfortunately, the author cannot obtain methylnaltrexone to compare results, and therefore cannot report the results of this personal experiment any further.

12.3 SEARCH FOR NEW PAINKILLERS

12.3.1 Painkillers From Black Mamba Venom

Acid-sensing ion channels (ASICs) are generally considered major players in the pain pathway, including in humans. A snake polypeptide toxin activating peripheral ASICs in nociceptive neurons has been shown to cause pain. Thus, the polypeptide toxins including the snake toxin play an important role in helping us to understand the physiological and physio-pathological functions of the ion channels. In the field of pain, the polypeptide toxins has caused important advances in basic research and even to clinical applications.

Diochot et al. (2012) found that a new three-finger peptide venom (*Dendroaspis polylepis polylepis*) from the black mamba was able to abolish pain through inhibition of ASICs expressed in either central or peripheral neurons. These peptides, called mambalgins (see Fig. 12.3.1 for mambalgin-2), are not toxic in mice but show potent analgesic effects on central and peripheral injection that can be as strong as morphine.

This effect is, however, resistant to naloxone (a medication used to block the effects of opioids), and mambalgins cause much less tolerance than morphine and no respiratory distress (RD), which is caused by morphine.

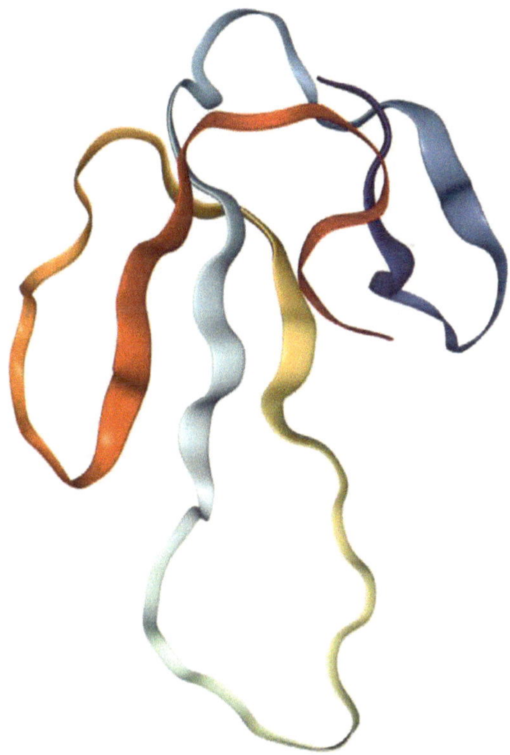

Fig. 12.3.1 Mambalgin-2. The structure is from Schroeder et al. (2014), which is 2MFA in RCSB Protein Data Bank.

Pharmacological inhibition by mambalgins combined with the use of knockdown and knockout animals indicates that blockade of heteromeric channels made of ASIC1a and ASIC2a subunits in central neurons and of ASIC1b-containing channels in nociceptors is involved in the analgesic effect of mambalgins. These findings identify new potential therapeutic targets for pain and introduce natural peptides that block them to produce a potent analgesia.

12.3.2 Centipede Venom

Voltage-gated sodium (Na_V) channels are integral transmembrane proteins that give a current pathway for the rapid depolarization of excitable cells. The Na_v channels play a key role in conveying nociceptor responses to synapses in the dorsal horn (see Fig. 12.1.1).

Humans contain nine different Na_V channel subtypes: $Na_V1.1$ to $Na_V1.9$. In recent years, $Na_V1.7$ has emerged as a promising analgesic target. Gain-of-function mutations in the SNC9A gene encoding the pore-forming α-subunit of $Na_V1.7$ cause severe episodic pain in inherited neuropathies, such as erythromelalgia and paroxysmal extreme pain disorder. In contrast, loss-of-function mutations in SCN9A result in a congenital indifference to pain (CIP). This can be recapitulated in rodents via complete knockout of $Na_V1.7$ in all sensory and sympathetic neurons. Moreover, certain polymorphisms in the SCN9A gene correlate with sensitivity to nociceptive inputs. Thus, the combined genetic data suggest that subtype-selective blockers of $Na_V1.7$ can be useful analgesics for treating a broad range of pain conditions.

Centipedes are one of the oldest existing arthropods, with a fossil record of 430 Myr. Centipedes were one of the first terrestrial taxa to use venom, and they have captured a wide variety of prey. The centipede venom apparatus has evolved by modification of the first pair of walking legs into a set of pincerlike claws (forcipules). Venom is secreted through a pore located near the tip of each forcipule. There are $\sim$3300 existent species of centipedes, yet the venom of only a small number has been studied in detail.

Undheim and King (2011) demonstrated that the venom of the Chinese red-headed centipede *Scolopendra subspinipes mutilans* is plenty with unique, disulfide-rich peptides that potently modulate the activity of mammalian voltage-gated ion channels (Yang et al., 2012). Thus Yang et al. (2013) decided to explore this venom as a potential source of $Na_V1.7$ inhibitors. They described the purification from this venom of a highly selective inhibitor of $Na_V1.7$, which is a more effective analgesic than morphine in rodent pain models.

Fig. 12.3.2 Solution structure of μ-SLPTX-Ssm6a (2MUN from RCSB Protein Data Bank) by Undheim et al. (2015).

In more detail, Yang et al. (2013) found μ-SLPTX-Ssm6a, which is a unique 46-residue peptide (see Fig. 12.3.2, obtained by Undheim et al., 2015) from the centipede venom that potently inhibits $Na_V1.7$ with an IC_{50} (half maximal (50%) inhibitory concentration) of ~25 nM. μ-SLPTX-Ssm6a has more than 150-fold selectivity for $Na_V1.7$ over all other human Na_V subtypes, except for $Na_V1.2$, for which the selectivity is 32-fold. Thus, μ-SLPTX-Ssm6a was a more potent analgesic than morphine in a rodent model of chemical-induced pain, and it was equipotent with morphine in rodent models of thermal and acid-induced pain. This study established μ-SPTX-Ssm6a as a promising molecule for the development of novel analgesics targeting $Na_V1.7$, which might be suitable for treating a wide range of human pain pathologies.

12.3.3 Computational Search for Analgesics Structure Without Side Effects

Morphine (see Fig. 12.1.2A) is an alkaloid from the opium poppy used to treat pain. The potentially lethal side effects of morphine and related opioids include fatal respiratory depression (RD). Recent studies have shown that opioid-induced analgesia results from μ-opioid-receptor (MOR) signaling through the G protein G_i, while many side effects, including RD and constipation, may be caused via β-arrestin pathway signaling downstream of MOR activation. Therefore, G protein MOR signaling alone is thought to cause only analgesia (see Fig. 12.3.3A).

Manglik et al. (2016) computationally put over 3 million molecules into the MOR structure pocket and found new scaffolds unrelated to known opioids. Seven molecules were finally selected. The structure-based optimization yielded PZM21 (see Fig. 12.3.3B)—a potent G_i activator with exceptional selectivity for MOR and minimal β-arrestin-2 recruitment. Unlike morphine, PZM21 is more effective for the affective component of analgesia versus the reflexive component, and is lack of both RD and morphine-like dose-escalation in mice at equi-analgesic doses. PZM21 thus serves as both a probe to disentangle MOR signaling and a therapeutic lead that is devoid of many of the side effects of current opioids.

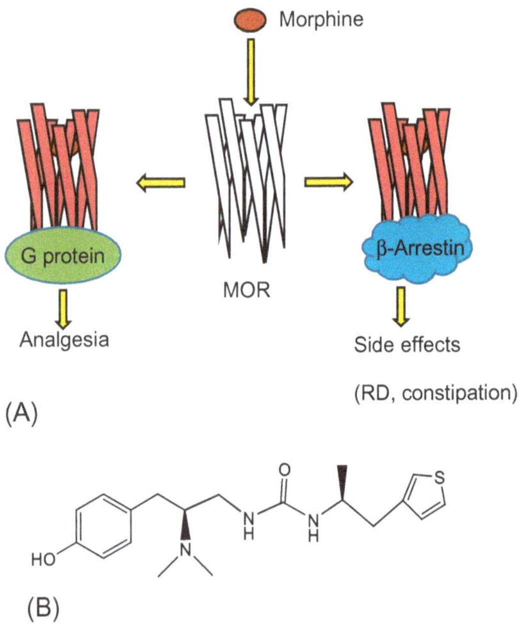

Fig. 12.3.3 (A) The μ-opioid-receptor (MOR) and MOR signaling. (B) a new analgesic. (A) When morphine (and other opioids) comes into μ-opioid-receptor (MOR), the MOR is activated. The activated MOR recruits G protein (G_i) or β-arrestin. The protein-MOR complex signals downstream, and finally cause analgesia or side effects (respiratory depression (RD), constipation, etc.), respectively. (B) The structure of PZM21.

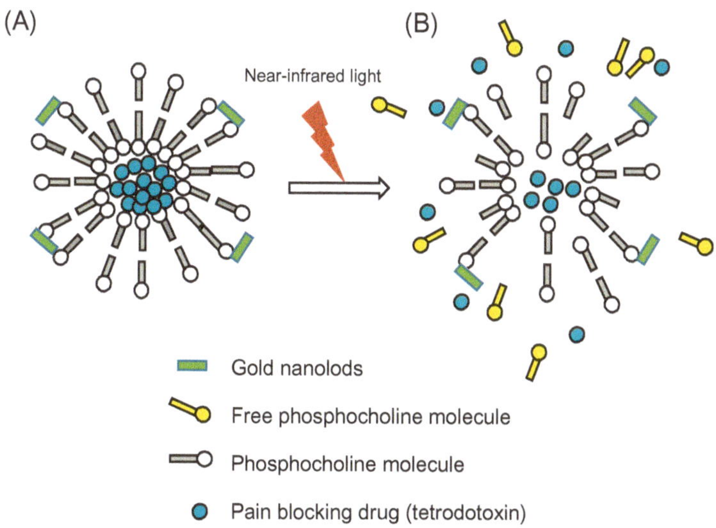

Fig. 12.4.1 On-demand pain relief, controlled by light. (A) Phospholipids are ready with pain blocking drug. (B) The phospholipids are broken by light, releasing the pain blocking drug.

12.4 ON-DEMAND PAIN RELIEF

Zhan et al. (2017) designed on-demand pain relief system, made of injectable liposomes, which is cell-like sacs consisted of phospholipids (see Fig. 12.4.1A). They filled liposome with the nerve blocker tetrodotoxin and attached gold nanorods to them.

The place where the anesthesia is required is heated by near-infrared light. The gold nanorods on the liposomes absorb heat efficiently, and the membrane expands, resulting in the drug leaking inside the liposomes (Fig. 12.4.1B). When the expanded liposomes become cool, they shrink again. This cycle can be repeated a few times.

12.5 PAIN RELIEF BY κ-OPIOID RECEPTOR (KOR)

Current opioids cause pain relief typically by binding to μ-opioid receptor on the nerve cell surface. However, these opioids cause side effects such as addiction. Thus, doctors are trying to use other member of the opioid receptors such as κ-opioid receptor (KOR) as a target of pain relief. Activating this receptor does not cause euphoria, which is a significant cause of opioid abuse. However, when the compounds bind to it, it could cause other unknown side effects such as sedation, dysphoria, depression, or a strong feeling of uneasiness.

Bohn and Aubé (2017) worked toward the discovery of biased agonists of the κ-opioid receptor (KOR) with high effects on G protein with high β-arrestin. These agonists cause dysphoria, etc. They found selective KOR agonist for an isoquinolinone hit and isoquinolinone 2.1 (see Fig. 12.5.1A and B). They also found advanced analogues: triazole probe and triazole 1.1 (see Fig. 12.5.1C and D).

Bohn and Aubé also discovered standard balanced KOR agonists with high effects on G protein with low β-arrestin, of the valuable heterocyclic compounds, U69,593 (U69), U50,488H (U50), and the nonnitrogenous natural product salvinorin A (see Fig. 12.5.1E–G, respectively).

12.6 OPIOID WITHOUT SIDE EFFECTS

A new opioid that works as a painkiller generally has negative side effects such as addiction, respiratory depression, and constipation. Opioid receptors out of the brain and spinal cord are more active at the sites of injury or inflammation. The pH is lower at such sites; therefore, more protons are around the sites. Thus, the opioids that work at lower pH would not affect the normal environment such as in the brain.

Based on this knowledge, Spahn et al. (2017) used computer simulations at low pH and designed an agonist that selectively activates peripheral μ-opioid receptors at the source of pain generation.

Unlike the conventional opioid fentanyl (see Fig. 12.6.1A), Spahn et al. made a fluorinated version of fentanyl, called NFEPP (see Fig. 12.6.1B). NFEPP has a pK_a of 6.8, so that it is protonated only in the low-pH condition of injured or inflamed tissue. In contrast, it is not protonated in the brain; therefore, it is inactive.

In rats, NFEPP produced injury-restricted analgesia with different types of inflammatory pain, without exhibiting respiratory depression, sedation, constipation, or addiction potential.

12.7 PAIN INDUCED BY SCORPION TOXIN

Venom has evolved to protect many defenseless animals with a lethal weapon. This effective survival strategy is demonstrated by scorpions, snakes, spiders, cone snails, and sea anemones (the "5S"s). Most animal

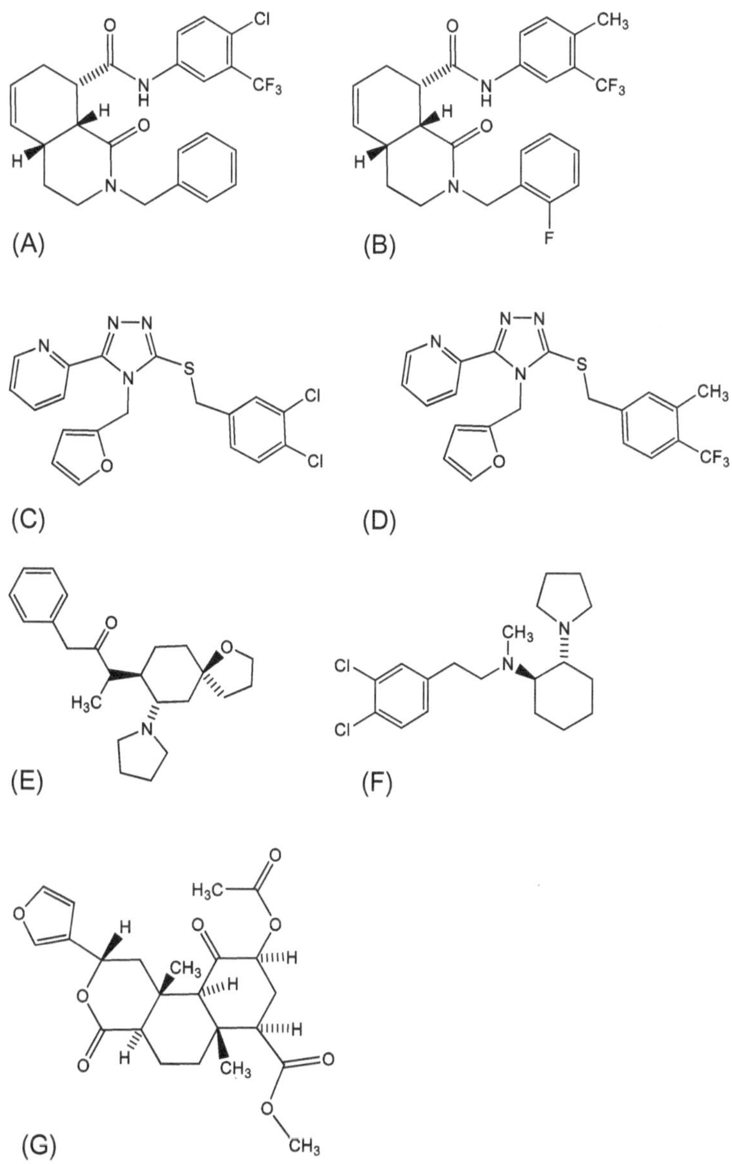

Fig. 12.5.1 Agonists for κ-opioid receptors. (A) Isoquinolinone hit. (B) Isoquinolinone 2.1. (C) Triazole probe. (D) Triazole 1.1. (E) U69,593 (U69). (F) U50,488H (U50). (G) Salvinorin A.

venoms contain multiple peptide toxins highly selective for their possible targets. Venomous animals use peptide toxins for hunting and self-defense. To achieve these goals, toxins need to bind to their targets with high affinity due to the small amount that a single bite or sting can deliver.

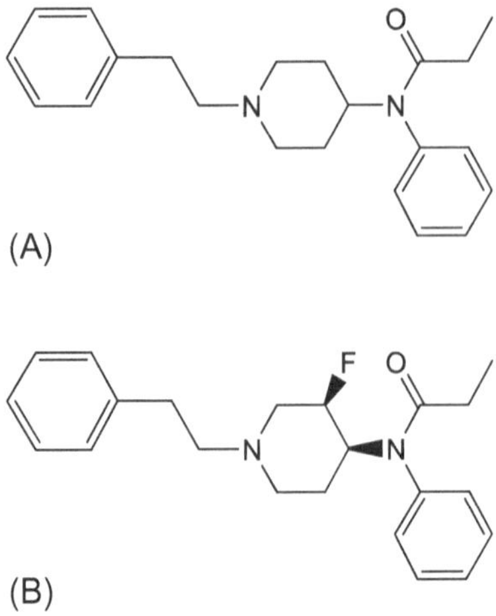

Fig. 12.6.1 (A) Fentanyl and (B) NFEPP.

Many scorpions can deliver lethal stings. These are known to be extremely painful—apparently enough to stop larger predators when they cannot be overpowered by venom toxicity alone. Like other venomous species, scorpions use a cocktail of toxins to manipulate multiple targets simultaneously. BmP01, a member of the inhibitor cysteine knot motif toxins with 29 amino acids (see Fig. 12.7.1 a for peptide sequence; see Fig.12.7.1B and C for structures after Wu et al. (2000)), is a small scorpion peptide toxin known to target the nociceptive TRPV1 ion channel (which is well known as the receptor for capsaicin) to produce pain (Hakim et al., 2015). The scorpion toxin BmP01 is related to sting-induced excruciating pain; however, the reported minimum concentrations for activating TRPV1 channel are micro-molar range, which appear too high to be affected by the toxin.

Yang et al. (2017) reported a mechanistic study of BmP01, which has revealed a new cooperative action between the toxin and protons, which are both from the scorpion venom, that gives high effectiveness in activating the toxin's target. They showed that the effective concentration of BmP01 is highly pH-dependent: it decreases more than 100-fold in activating TRPV1 (see Fig. 12.7.1D). When the TRPV1 is open, we feel pain. The TRPV1 channel is highly dependent on pH and BmP01. BmP01 binds to one of the two proton-binding sites on TRPV1 and, together with a proton, strongly activates the nociceptive channel. Because most animal venoms are acidic, proton-facilitated synergistic action may represent a general strategy for maximizing toxin potency.

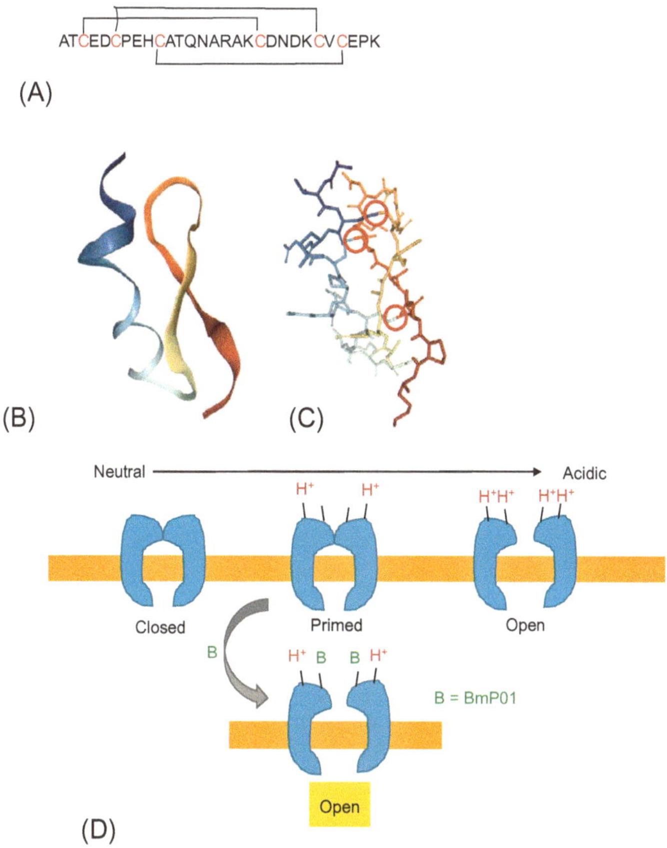

Fig. 12.7.1 (A) The peptide sequence of the scorpion toxin, BmP01, and disulfide bonds (C is cysteine; see Table 1 for others). (B) Structure of BmP01 in the ribbon style (PDB: 1WM7). (C) BmP01 in the licorice style. The circles in panel C show the disulfide bonds. (D) TRPV1 channel. See text for details.

12.8 A HIGHER PRICE TAG CAUSES STRONGER NOCEBO HYPERALGESIA

Value information about a drug, such as the price tag or the package, could strongly affect its therapeutic effect. Tinnermann et al. (2017) discovered that value information influences adverse treatment outcomes in humans without an active substance. Labeling an inert treatment as expensive medication led to stronger nocebo hyperalgesia (feeling more pain) than labeling it as cheap medication.

Tinnermann et al. (2017) found this effect was caused by neural interactions among the cortex, brainstem, and spinal cord, using fMRI. Activity in the prefrontal cortex mediated the effect of value on nocebo hyperalgesia. Value furthermore modulated coupling between prefrontal areas, brainstem, and spinal cord, which might represent a flexible mechanism through which higher-cognitive representations, such as value, can modulate early pain processing.

Tinnermann et al. (2017) showed that expensive medication increases the risk for developing nocebo-related side effects. Moreover, their fMRI protocol made it possible to assess how value information about a medical treatment affects the entire central human pain system. Two regions of the descending pain pathway, the rACC and the PAG, helped expectation-induced pain modulation and transported the difference in nocebo effects between cheap and expensive treatment.

REFERENCES

Bohn, L.M., Aubé, J., 2017. Seeking (and Finding) Biased Ligands of the Kappa Opioid Receptor. ACS Med. Chem. Lett. 8, 694–700.

Corder, G., Tawfik, V.L., Wang, D., Sypek, E.I., Low, S.A., et al., 2017. Loss of μ opioid receptor signaling in nociceptors, but not microglia, abrogates morphine tolerance without disrupting analgesia. Nat. Med. 23, 164–173.

Diochot, S., Baron, A., Salinas, M., Douguet, D., Scarzello, S., et al., 2012. Black mamba venom peptides target acid-sensing ion channels to abolish pain. Nature 490, 552–555.

Hakim, M.A., Jiang, W., Luo, L., Li, B., Yang, S., et al., 2015. Scorpion toxin, BmP01, induces pain by targeting TRPV1 channel. Toxins 7, 3671–3687.

Manglik, A., Lin, H., Aryal, D.K., McCorvy, J.D., Dengler, D., et al., 2016. Structure-based discovery of opioid analgesics with reduced side effects. Nature 537, 185–190.

Schroeder, C.I., Rash, L.D., Vila-Farres, X., Rosengren, K.J., et al., 2014. Chemical synthesis, 3D structure, and ASIC binding site of the toxin mambalgin-2. Angew. Chem. Int. Ed. Engl. 53, 1017–1020.

Spahn, V., Del Vecchio, G., Labuz, D., Rodriguez-Gaztelumendi, A., Massaly, N., et al., 2017. A nontoxic pain killer designed by modeling of pathological receptor conformations. Science 355, 966–969.

Thomas, J., Karver, S., Cooney, G.A., Chamberlain, B.H., Watt, C.K., 2008. Methylnaltrexone for opioid-induced constipation in advanced illness. New Engl. J. Med. 358, 2332–2343.

Tinnermann, A., Geuter, S., Sprenger, C., Finsterbusch, J., Büchel, C., 2017. Interactions between brain and spinal cord mediate value effects in nocebo hyperalgesia. Science 358, 105–108.

Undheim, E.A., King, G.F., 2011. On the venom system of centipedes (Chilopoda), a neglected group of venomous animals. Toxican 57, 512–524.

Undheim, E.A., Grimm, B., Low, L.L., Morgenstern, C.F., Herzig, D.V., et al., 2015. Solution structure of μ-SLPTX3-Ssm6a. Structure 23, 1283–1292.

Wu, G., Li, Y., Wei, D., He, F., Jiang, S., et al., 2000. Solution structure of BmP01 from the venom of scorpion Buthus martensii Karsch. Biochem. Biophys. Res. Commun. 276, 1148–1154.

Yang, S., Liu, Z., Xiao, Y., Rong, M., Liang, S., et al., 2012. Chemical punch packed in venoms makes centipedes excellent predators. Mol. Cell. Proteomics 11, 640–650.
Yang, S., Xiao, Y., Kang, D., Liu, J., Li, Y., et al., 2013. Discovery of a selective $Na_V1.7$ inhibitor from centipede venom with analgesic efficacy exceeding morphine in rodent pain models. 2013. PNAS 110, 17534–17539.
Yang, S., Yang, F., Zhang, B., Lee, B.H., Li, B., et al., 2017. A bimodal activation mechanism underlies scorpion toxin–induced pain. Sci. Adv. 3 (8), e1700810.
Zhan, C., Wan, W., Santamaria, C., Wang, B., Rwei, A., et al., 2017. Ultrasensitive phototriggered local anesthesia. Nano Lett. 17, 660–665.

FURTHER READING

McMahon, S.B., Koltzenburg, M., Tracey, I., Turk, D.C., 2013. In: McMahon, S.B. et al., (Ed.), Wall and Melzack's Textbook of Pain, sixth ed. Elsevier Saunders, Philadelphia, p. 1153.
Nagase, H., Fujii, H., 2010. Opioids in preclinical and clinical rials. In: Nagase, H. (Ed.), Chemistry of Opioids. Topics in Current Chemistry. vol. 299. Springer, Berlin, Heidelberg.

CHAPTER THIRTEEN

Disease-Related Topics and Materials

Contents

13.1 NEW POLYMERS FOR PROTECTING INSULIN

Protein drugs have high selectivity and potential—more than 130 proteins or peptides have been approved by the FDA. However, proteins have not only short half-lives in the bloodstream, but also short shelf-lives because of their chemical and physical instability.

The polymer conjugation technique as poly(ethylene glycol) (PEG) or PEGylation to proteins is the most popular method, and 10 FDA-approved PEGylated proteins are on the market at the time of writing. However, linear PEG does not improve the stability during storage and transport. Nearly 80% of protein drugs need to be refrigerated or frozen even with PEG

https://doi.org/10.1016/B978-0-12-817054-0.00013-8
© 2019 Elsevier Inc.
All rights reserved.

(Leader et al., 2008). This lowers quality of life of the proteins and increases costs due to refrigeration in delivery and storage. Furthermore, degradation of the drugs causes inadequate dosage.

Many replacing methods for the PGE methods have been used to lengthen the half-life of protein, but few can increase stability. Protein-polymer conjugates, such as poly(N-(2-hydroxypropyl) methacrylamide) (pHPMA), polyoxazolines, and hydroxyethyl starch (HES), successfully increased the length of life in vivo. However, these polymers do not improve protein stability during storage.

There are a few recent examples of protein-stabilizing polymers, such as zwitterionic carboxybetaine polymer for enzymes (Keefe and Jiang, 2011), cationic dendrimer for proline-specific endopeptidase (PEP) (Fuhrmann et al., 2013), and branched or "comb" PEG for oxytocin (Collins et al., 2015). However, only enzymes or small peptides are targeted. A great need and interest exist for developing new PEG alternatives that have longer half-lives with storage stability.

Liu et al. (2017) have developed a trehalose glycopolymer named Poly-Protek, which increases protein stability during both storage and use. The polymer has the disaccharide trehalose at the side chains and stabilizes various enzymes and proteins to variations of temperature and lyophilization in solution.

Liu et al. (2017) used a trehalose glycopolymer as a resisting material. They demonstrated that the polymer stabilizes proteins in a high vacuum. Trehalose is well-known for stabilizing proteins with a mechanism attributed to vitrification, water replacement, and entrapment.

Trehalose glycopolymers have proven to be better than trehalose itself in several studies. This may be a result of the surfactant character of the polymers. Liu et al. (2017) hypothesized that trehalose glycopolymers would also stabilize proteins in vivo, similar to PEG. They demonstrated for the first time that the trehalose glycopolymer (see Fig. 13.1.1A) can stabilize a therapeutic protein at high temperatures and under mechanical stress.

Liu et al. chose insulin (see Fig. 13.1.1B; PDB 4INS; Baker et al., 1988) as a model protein because of its wide clinical usage and well-established structure and bioactivity assays. The insulin-trehalose glycopolymer conjugate (see Fig. 13.1.1B) retained bioactivity far longer even after exposure to high temperatures (90°C, 30 min) and mechanical agitation for 3 h. It showed no toxicity, and significantly longer plasma circulation time in mice.

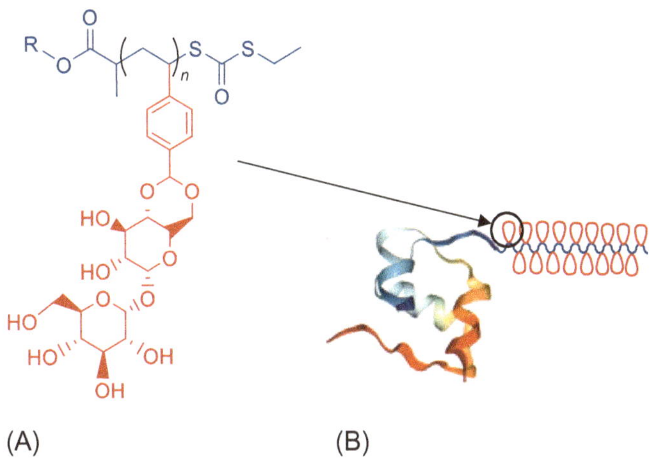

Fig. 13.1.1 (A) Trehalose glycopolymer. (B) Insulin-trehalose glycopolymer conjugate. Insulin is from PDB, 4INS.

13.2 ARTIFICIAL VESICLE FOR REGULATING GLUCOSE LEVELS IN MICE

Diabetes is a chronic disease characterized by disorder of glucose regulation. Insulin is a peptide hormone, which regulates blood glucose levels. Type I diabetes patients require daily injections of insulin.

Xiao et al. (2017) made a new polymer vesicle (polymersome) to regulate glucose levels in diabetes patients not by insulin but by a nanosized "sugar sponge." This is a lectin-bound glycopolymersome capable of regulating glucose via the dynamic recognition between the lectin and different carbohydrates.

The glycopolymersome is self-assembled from poly(ethylene oxide)-*block*-poly[(7-(2-methacryloyloxyethoxy)-4-methylcoumarin)-*stat*-2-(diethylamino)ethyl methacrylate-*stat*-(α-D-glucopyranosyl)ethyl methacrylate] [PEO_{43}-b-P(CMA_8-*stat*-DEA_{80}-*stat*-$GEMA_{12}$)] (see Fig. 13.2.1A). In this context, "*stat*" means statistically—for example, CDG or CGD or DCG, etc., but stoichiometry of C:D:G is kept as 1:1:1.

Then the Con A, which is a lectin with high affinity for glucose and mannose, is added. The lectin bound in the glycopolymersome has a different affinity for the glucose in the blood and the glucosyl group in the glycopolymersome. Therefore, this sugar sponge works as a glucose storage unit

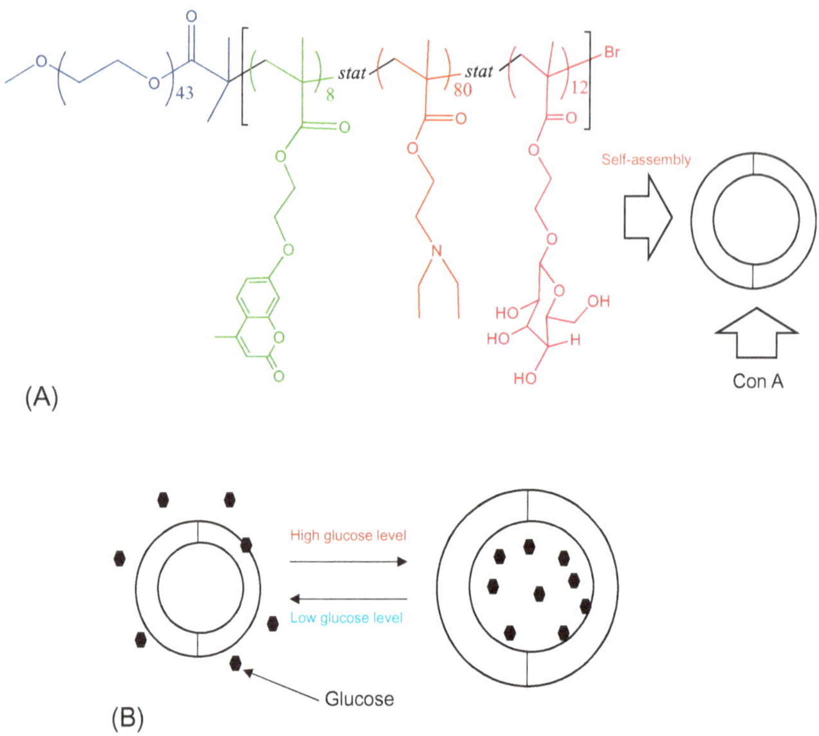

Fig. 13.2.1 Glycopolymersome for regulating blood glucose levels. (A) PEO_{43}-b-P(CMA_8-*stat*-DEA_{80}-*stat*-$GEMA_{12}$), and self-assembly into glycopolymersome. Then a lectin (Con A) is added. (B) Glucose levels are kept to be constant as if the Con A breathes the glucose.

by dynamic sugar replacement. The lectin in the sugar sponge binds and stores the glucose from its surrounding solution when the glucose concentration is too high. The lectin releases the glucose when the glucose concentration is too low (see Fig. 13.2.1B).

In vitro, this sugar-breathing behavior is reflected to a remarkable size change of the sugar sponge due to the swelling or shrinkage at high or low glucose levels, respectively, which can be employed for blood sugar monitoring.

In vivo, this sugar sponge showed an excellent antidiabetic effect for type I diabetic mice within 2 days of one dose, which is much longer than traditional long-acting insulin.

In conclusion, this concept of "controlling sugar levels with sugar" is a totally new idea for regulating the blood glucose level without the involvement of insulin or other antidiabetic drugs after careful testing in humans.

13.3 RELATIONSHIP BETWEEN TYPE 1 DIABETES (T1D) AND GUT MICROBIOTA

13.3.1 Type 1 Diabetes (T1D)

Type 1 diabetes (T1D) is an autoimmune disease. T1D comes from the specific destruction of insulin-producing β-cells in the pancreas. Although T1D has a strong genetic basis, the incidence of T1D has been increasing in recent decades, which indicates a strong rising contribution from the environment. Environmental influences may result from hygiene, antibiotic use, and diet. Many of these factors have a relation to the "gut microbiome" in some way.

Similar to T1D in humans, diabetes in the nonobese (nonfat) diabetic (NOD) mouse model comes from the autoimmune destruction of pancreatic β-cells. The NOD mice lack the inborn signaling molecule of MyD88. The mice are protected from the development of T1D. However, such protection is lost when $Myd88^{-/-}$ mice of the NOD strain (NOD.*Myd88*$^{-/-}$ mice) are housed in germ-free conditions. The absence of MyD88 in NOD mice results in over-representation of the bacterial phylum Bacteroidetes and suppression of diabetes, probably due to production of an immunomodulatory microbial material. The higher occurrence of diabetes in female NOD mice than in male NOD mice is also ascribed to the effects of gut microbiota.

13.3.2 Importance of Short-Chain Fatty Acids (SCFAs) Produced by Gut Microbiota

Recently, influences of gut microbiota and immune responses by molecular mechanisms have attracted great attention from researchers. The main metabolites of the gut microbiota are short-chain fatty acids (SCFAs). Large amounts of SCFAs are produced in the colon through bacterial fermentation of dietary fiber. SCFAs raise the number and function of induced regulatory T cells (T_{reg} cells) in the colon. Furthermore, acetate elevates gut barrier function, and butyrate has antiinflammatory effects by inhibition of a transcription factor NF-κB. SCFAs and other metabolites bind to "metabolite-sensing" G-protein-coupled receptors (GPCRs; see Section 4.2.1). Mice deficient in one such receptor, GPR43, worsen inflammation in models of colitis and peripheral inflammation. SCFAs also inhibit histone-deacetylase (HDAC) activity and thereby influence gene transcription, including that in T_{reg} cells.

13.3.3 Effects of SCFAs on T1D

The T1D is a T cell-related disease and related to reduction of Foxp3$^+$ T_{reg} cells. Both the patients of T1D and the NOD mice displayed damaged gut barrier function. Changes of the composition of the gut microbiota are associated with many inflammatory diseases, such as T1D. In the gut, SCFAs are important for maintaining immunological tolerance to food antigens.

13.3.4 Approach Against T1D by Foods Using Bacterial Metabolites

Mariño et al. (2017) used diets that produce large amounts of acetate or butyrate to determine the individual effects and the combined effects of SCFAs on the development of diabetes in the NOD mice. They found new mechanisms of SCFA. Their approach is nonpharmaceutical, instead using diets and bacterial metabolites. Such an approach is effective in the prevention of autoimmune diseases such as T1D.

In more detail, Mariño et al. (2017) fed the NOD mice High-Amylose Maize Starch (HAMS), which is either acetylated (HAMSA) or butylated (HAMSB). This starch is intended to produce large amounts of acetate or butyrate after bacterial fermentation in the colon. They also fed the mice special diets that release large amounts of specific SCFAs after bacterial fermentation. A high degree of protection from diabetes was observed after ingestion of these foods. Complete protection was obtained by a combined acetate- and butyrate-yielding diet.

Acetate remarkably decreased the frequency of autoreactive T cells in lymphoid tissues, through effects on B cells and their ability to expand populations of autoreactive T cells. Foods containing butyrate increased the number and function of regulatory T cells. Both acetate- and butyrate-yielding foods (HAMSA and HAMSB) enhanced gut integrity and decreased the concentration of diabetogenic cytokines such as IL-21 in serum. Medicinal foods may be an effective and natural approach for countering the numerous immunological defects that contribute to T cell-dependent autoimmune diseases.

For humans, the cause of T1D is not clear. However, a large amount of acetate and/or butyrate could be an effective therapy. A medicinal food approach is very important for T1D, because this disease usually arises in children and young adults (adults are not considered because of the general strength to diseases), who need maximum drug-safety. HAMSA and HAMSB supplements have already been used successfully in clinical trials for gastrointestinal tract disorders (Le Leu et al., 2015). The studies of Mariño et al. (2017) also emphasize the possibilities of medicinal foods for the treatments of human diseases.

13.4 STUDIES OF IMMUNE CELLS

13.4.1 Quantification of Communication Distances of Immune Cells

Immune cells communicate with each other by exchanging cytokines to achieve an appropriate response. However, the distances over which such communication occurs are not known.

Oyler-Yaniv et al. (2017) used theoretical and experimental models of immune responses to quantify the spatial extent of cytokine communications. They established that competition between cytokine diffusion and consumption gave spatial niches, λ_{niche} $(= \sqrt{[D/\ (n_{\text{consumers}}\ k_{\text{consumption}})]})$, of high cytokine concentrations with sharp boundaries. The sizes of these self-assembled niches scaled with the density of cytokine-consuming cells, a parameter that is tuned during immune responses.

In vivo, they measured λ_{niche} on length scales of 80–120 mm, which resulted in a high degree of cell-to-cell variance in cytokine exposure. Such heterogeneous distributions of cytokines were a source of nongenetic cell-to-cell variability that is often overlooked in single-cell studies. Their findings thus provide a basis for understanding variability in the patterning of immune responses by diffusible factors.

The diffusion factors are determined by the microfluidic device described in Section 4.4.

13.4.2 Sunscreen Blocks Autoimmune Encephalomyelitis (EAE) in Mice

UV light suppresses experimental autoimmune encephalomyelitis (EAE), a widely used animal model of multiple sclerosis (MS) in mice. UV light may be responsible for the decreased incidence of MS in equatorial regions.

To test this concept in more detail, Wang et al. (2017) applied commercially available sunscreen preparations to mice before exposing them to UV radiation. Surprisingly, some of the sunblock preparations blocked EAE without UV radiation. Furthermore, various sunblock preparations had variable abilities to suppress EAE. Wang et al. (2017) identified that the most effective agents for suppressing EAE were homosalate (Fig. 13.4.1A) and

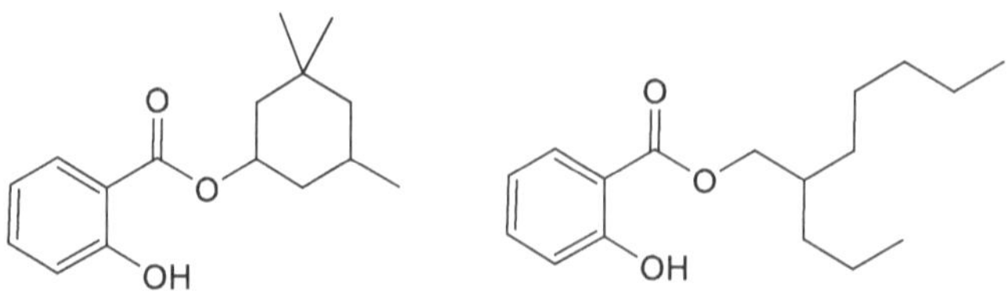

(A) (B)

Fig. 13.4.1 Sunscreens: (A) homosalate. (B) octisalate.

octisalate (Fig. 13.4.1B). Thus, salates should be useful in stopping the progression of MS, and may provide new insight into mechanisms of controlling autoimmune disease.

13.4.3 A Subtype of Immune Cells Associated With Allergies Were Found

Although T helper type 2 (T_H2) cells give necessary protection to some pathogens, they are also associated with allergy pathogenesis. However, there have been no methods to distinguish the pathogenic cells that are reactive to allergens from the rest of the T_H2 population.

Wambre et al. (2017) have described a subset of human memory T_H2 cells that includes all allergen-specific T_H2 cells. These cells are differentiated $CD4^+$ T cells ($CD27^-$ and $CD45RB^-$) characterized by coexpression of CRT_H2, CD49d, and CD161; they also exhibit numerous functional attributes distinct from conventional T_H2 cells. Thus, Wambre et al. have called these cells of allergic disease-related phenotype the T_H2A cell subset.

These cells decreased in patients after allergen immunotherapy, indicating that targeting T_H2A cells could reduce allergic responses. These findings identify a human T_H2 cell signature in allergy-related diseases that could be used for response-monitoring and designing appropriate immunomodulatory strategies.

13.5 A NEW ROLE OF ATP AS A HYDROTROPE

Adenosine triphosphate (ATP; see Fig. 13.5.1) carries energy in a cell, serves as a signal molecule, and is a source of adenosine in RNA. As shown in the figure, ATP is composed of two parts: hydrophilic and hydrophobic.

Patel et al. (2017) have found a new function of ATP, which works as a biological hydrotrope in a cell. Hydrotropes are small molecules that solubilize molecules in aqueous solutions. The hydrotrope is different from the

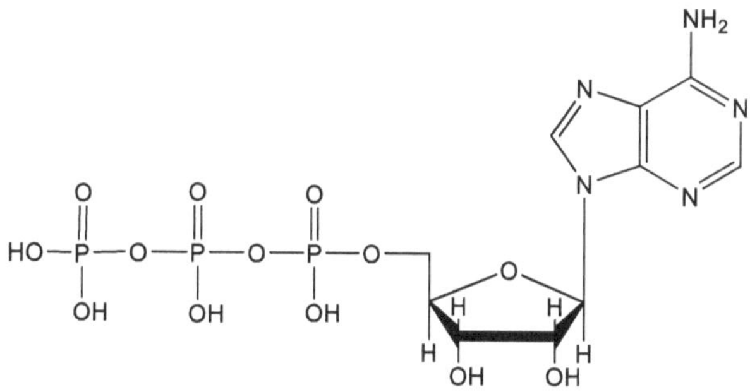

Fig. 13.5.1 ATP.

surfactants because the surfactants do not aggregate each other so that they do not form micelles.

Patel et al. made 1–20 mM ATP with Mg^{2+} (ATP-Mg). Even for 8 mM ATP-Mg, they did not carry out liquid-liquid separation. They used four proteins—FUS-GFP, TAF15-Snap-Alexa Fluor 546, hnRNPA3-GFP and PGL-3-GFP—all of which easily aggregate and separate in the cell. They mixed ATP-Mg and each protein, and found that 2–8 mM of ATP-Mg is enough to inhibit the phase separation of each protein.

Furthermore, they added crude egg white to 10 mM ATP-Mg, and found that the mixture heated at 60°C for 30 min aggregated <10%, while the egg white without ATP-Mg aggregated >95% in 5 min.

In conclusion, ATP keeps proteins soluble in cytoplasm. Patel et al. also suggested that ATP could eventually become the basic energy source for metabolism, which is the hydrolysis of ATP to ADP.

13.6 DIFFERENT TYPES OF ALZHEIMER'S DISEASE

The accumulation of amyloid-β fibrils in the brain is the most important feature of Alzheimer's disease. The amyloid-β fibrils are polymorphic, and, there are two variations in molecular structure in vitro, Aβ40 and Aβ42, of which the differences are amyloid-β peptides of 40- and 42-residue, depending on fibril growth conditions.

Recent experiments (e.g., Petkova et al., 2005) showed that variations in amyloid-β fibril structure in vivo were associated with variations in the Alzheimer's disease phenotype. This is similar to prion strains being correlated with different pathological phenotypes.

Qiang et al. (2017) explored the correlation between the structural variation and the Alzheimer's disease phenotype using solid-state nuclear magnetic resonance (ssNMR) measurements on Aβ40 and Aβ42 fibrils grown from extracts of brain cortex with Alzheimer's disease.

Qiang et al. (2017) compared a typical prolonged-duration form (t-AD) with two atypical subtypes; one is the rapidly progressive form (r-AD) and the other is the posterior cortical atrophy type (PCA-AD), using a two-dimensional (2-D) plot of ^{15}N-^{13}C NMR spectra.

The 37 instances of ssNMR data from 18 individuals have shown that the Aβ40 fibril structure is most abundant in t-AD and PCA-AD, while Aβ40 fibrils from r-AD exhibit a large portion of additional structures. Data from Aβ42 fibrils indicate structural heterogeneity in most samples from all patient categories, with at least two dominant structures.

These results indicate the existence of a specific predominant Aβ40 fibril structure in t-AD and PCA-AD, suggesting that r-AD may relate to additional fibril structures. The results also indicate that there is a qualitative difference between Aβ40 and Aβ42 aggregates in the brain tissue of patients with Alzheimer's disease.

13.7 THE ROLE OF THE POLYGLUTAMINE TRACT

A polyglutamine (polyQ) tract is a portion formed by a sequence of several repeated glutamine units in a protein. Typically, a polyglutamine tract consists of 10 to a few hundreds of the units.

Nine neurodegenerative diseases are caused by polyQ tracts in different proteins, such as huntingtin in Huntington's disease and ataxin 3 in spinocerebellar ataxia type 3 (SCA3).

The age of onset for such disease decreases with increasing polyQ length in these proteins, and the normal length also varies. PolyQ causes these diseases because isolated polyQ tracts are toxic, and an N-terminal huntingtin fragment comprising exon 1, which occurs in vivo as a result of alternative splicing, causes toxicity. ("Exon" is parts of mRNA removed by RNA splicing; "RNA splicing" is mRNA cutting into intron and exon to make the final form of mRNA; "introns" are not used in the final form of mRNA and exons are used; and "alternative splicing" is the splicing that cut out various parts of introns to make various mRNA, and exons are produced.) Although such mutant proteins tend to accumulate (Rubinsztein, 2006), toxicity is caused by soluble forms of the proteins. The function of the polyQ tracts in many normal cytoplasmic proteins has not yet been established.

One such protein is the deubiquitinating enzyme ataxin 3, which is widely expressed in the brain. Ashkenazi et al. (2017) show that the polyQ domain enables wild-type ataxin 3 to interact with beclin 1, a key initiator of autophagy, which is a destructive mechanism of the cell to disassemble unnecessary components. This interaction allows the deubiquitinase activity of ataxin 3 to protect beclin 1 from proteasome-mediated degradation ("proteasome" is a protein complex which degrades unneeded proteins) and thereby enables autophagy.

Starvation-induced autophagy, which is regulated by beclin 1, was particularly inhibited in ataxin3-depleted human cell lines and mouse primary neurons, and in vivo in mice. This activity of ataxin 3 and its polyQ-mediated interaction with beclin 1 was competed for by other soluble proteins with polyQ tracts in a length-dependent fashion. This competition resulted in the prevention of starvation-induced autophagy in cells expressing mutant huntingtin exon 1, and this impairment was repeated in the brains of a mouse model with Huntington's disease and in cells from patients.

A similar phenomenon also appeared with other polyQ disease proteins, including mutant ataxin 3 itself. Thus Ashkenazi et al. (2017) have described a specific function for a wild-type polyQ tract that is aborted by a competing longer polyQ mutation in a disease protein, and identified a harmful function of such mutations distinct from their tendency to aggregate.

REFERENCES

Ashkenazi, A., Bento, C.F., Ricketts, T., Vicinanza, M., Siddiqi, F., et al., 2017. Polyglutamine tracts regulate beclin 1-dependent autophagy. Nature 545, 108–111.

Baker, E.N., Blundell, T.L., Cutfield, J.F., Cutfield, S.M., Dodson, E.J., et al., 1988. The structure of 2Zn pig insulin crystals at 1.5 a resolution. Philos. Trans. R. Soc. London Ser. B 319, 369–456.

Collins, J., Tanaka, J., Wilson, P., Kempe, K., Davis, M.H., et al., 2015. In situ conjugation of Dithiophenol Maleimide polymers and oxytocin for stable reversible polymer-peptide conjugates. Bioconjug. Chem. 26, 633–638.

Fuhrmann, G., Grotzky, A., Lukic, R., Matoori, S., Luciani, P., et al., 2013. Sustained gastrointestinal activity of dendronized polymer-enzyme conjugates. Nat. Chem. 5, 582–589.

Keefe, A.J., Jiang, S., 2011. Poly(zwitterionic)protein conjugates offer increased stability without sacrificing binding affinity or bioactivity. Nat. Chem. 4, 59–63.

Leader, B., Baca, Q.J., Golan, D.E., 2008. Protein therapeutics: a summary and pharmacological classification. Nat. Rev. Drug Discov. 7, 21–39.

Le Leu, R.K., et al., 2015. Butyrylated starch intake can prevent red meat-induced O6-methyl-2-deoxyguanosine adducts in human rectal tissue: a randomised clinical trial. Br. J. Nutr. 114, 220–230.

Liu, Y., Lee, J., Mansfield, K.M., Ko, J.H., Sallam, S., et al., 2017. Trehalose glycopolymer enhances both solution stability and pharmacokinetics of a therapeutic protein. Bioconjug. Chem. 28, 836–845.

Mariño, E., Richards, J.L., McLeod, K.H., Stanley, D., Yap, Y.A., et al., 2017. Gut microbial metabolites limit the frequency of autoimmune T cells and protect against type 1 diabetes. Nat. Immun. 5, 552–562.
Oyler-Yaniv, A., Oyler-Yaniv, J., Whitlock, B.M., Liu, Z., Germain, R.N., et al., 2017. A tunable diffusion-consumption mechanism of cytokine propagation enables plasticity in cell-to-cell communication in the immune system. Immunity 46, 609–620.
Patel, A., Malinovska, L., Saha, S., Wang, J., Alberti, S., et al., 2017. ATP as a biological hydrotrope. Science 356, 753–756.
Petkova, A.T., Leapman, R.D., Guo, Z., Yau, W.M., Mattson, M.P., et al., 2005. Self-propagating, molecular-level polymorphism in Alzheimer's β-amyloid fibrils. Science 307, 262–265.
Qiang, W., Yau, W.-M., Lu, J.-X., Collinge, J., Tycko, R., 2017. Structural variation in amyloid-β fibrils from Alzheimer's disease clinical subtypes. Nature 541, 217–221.
Rubinsztein, D.C., 2006. The roles of intracellular protein-degradation pathways in neurodegeneration. Nature 443, 780–786.
Wambre, E., Bajzik, V., DeLong, J.H., O'Brien, K., Nguyen, Q.-A., et al., 2017. A phenotypically and functionally distinct human T_H2 cell subpopulation is associated with allergic disorders. Sci. Transl. Med. 9 (401). eaam9171.
Wang, Y., Marling, S.J., Plum, L.A., DeLuca, H.F., 2017. Salate derivatives found in sunscreens block experimental autoimmune encephalomyelitis in mice. PNAS 114, 8528–8531.
Xiao, Y., Sun, H., Du, J., 2017. Sugar-breathing Glycopolymersomes for regulating glucose level. J. Am. Chem. Soc. 139, 7640–7647.

CHAPTER FOURTEEN

Therapy-Related Topics

Contents

14.1 MAGNESIUM SILICIDE NANOPARTICLES FOR CANCER STARVATION THERAPY

An oxygen scavenger or deoxygenation agent (DOA) is a material that rapidly absorbs molecular oxygen. DOA has various industrial applications, such as in food preservation and anticorrosion of metal. As oxygen is necessary to cancer growth, DOA can be used as a possible cancer therapy, where the tumors are starved of oxygen by the intratumoral consumption of oxygen.

Zhang et al. (2017) made an injectable polymer-modified magnesium silicide (Mg_2Si) nanoparticle. They found that it works as a DOA for scavenging oxygen in a tumor when injected directly into the tumor, and the by-products prevent tumor blood capillaries from being deoxygenated. Thus, Zhang et al. (2017) have developed a new therapy against cancer.

The 100 nm diameter Mg_2Si nanoparticles are synthesized by modification with poly(vinylpyrrolidone) at high temperatures. As the tumor microenvironment is acidic, the injected Mg_2Si nanoparticle breaks and releases silane (SiH_4). This reacts efficiently with both tissue-dissolved and hemoglobin-bound oxygen to form silicon oxide (SiO_2) aggregates. This SiO_2 condensate, formed in situ, blocks the tumor blood capillaries, resulting in a lack of new supplies of oxygen and nutrients for any tumors.

Biochemistry for Materials Science
https://doi.org/10.1016/B978-0-12-817054-0.00014-X
© 2019 Elsevier Inc.
All rights reserved.

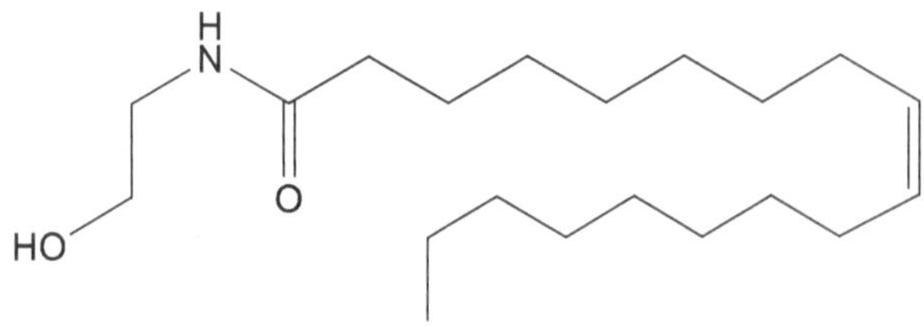

Fig. 14.2.1 Oleylethanolamide (OEA).

14.2 MOLECULAR MECHANISM BEHIND SUCCESS OF GASTRIC BYPASS SURGERY

The most successful weight loss therapy for obesity is gastric bypass surgery. Hankir et al. (2017) have identified a molecular pathway between the gut and the brain, which may underlie the success of the surgery.

Gastric bypass surgery involves removal of most of the stomach and reconnecting the remained small stomach and intestine, which makes patients feel full much faster. When Hankir et al. (2017) performed the surgery on rats, the rats not only lost weight, but also preferred low-fat meals to high-fat meals.

Furthermore, Hankir et al. (2017) found that the surgery increased synthesis of oleoylethanolamide (OEA; see Fig. 14.2.1) in the intestines. They believed that this change caused the loss of weight as well as the alteration in the rats' behavior. OEA is a lipid that binds α-type peroxisome proliferator-activated receptor (PPAR-α). PPAR-α controls absorbing, storing, and using fat. When PPAR-α was chemically inhibited in the rats that had the bypass surgery, their preference changed from low-fat meals to high-fat ones.

Moreover, Hankir et al. (2017) showed that the OEA connects the gut to the brain through the vagus nerve because the rats reverted to preferring high-fat meals when the nerve was cut or downstream dopamine receptors in the brain were blocked.

14.3 UTILIZING CANCER TISSUE STIFFNESS AS A DRUG TARGET

In patients, tumor cells grow by interactions with their microenvironment. A key part of the microenvironment is the extracellular matrix, which typically has a greater stiffness in tumors than in surrounding normal tissues.

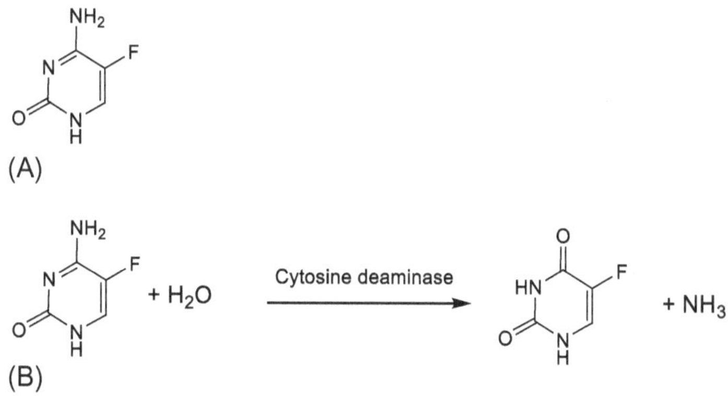

Fig. 14.3.1 (A) 5-Fluorocytosine. (B) Synthesis of 5-fluorouracil by cytosine deaminase.

Data suggest a strong correlation between collagen cross-linking and increased tissue stiffness at metastatic sites.

To take advantage of this, Liu et al. (2017) engineered a mechanoresponsive cell system (MRCS) to act as a vehicle for cancer drug delivery. MRCS uses mechanosensitive promoter-driven mesenchymal stem cell (MSC)–based vectors, which selectively home in on and target cancer metastases in response to specific mechanical cues. They deliver therapeutics to effectively kill cancer cells, as demonstrated in a metastatic breast cancer mouse model. These MRCS cells, accumulated in tumors, delivered the first half of a two-part cancer therapy: the enzyme cytosine deaminase. A drug called 5-fluorocytosine (see Fig. 14.3.1A) was then delivered systemically, and cytosine deaminase (see Fig. 14.3.1B) in the tumors activated the drug (5-fluorouracil; see Fig. 14.3.1B), providing local anticancer activity with no off-target damage in mice.

REFERENCES

Hankir, M.K., Seyfried, F., Hintschich, C.A., Diep, T.-A., Kleberg, K., et al., 2017. Gastric bypass surgery recruits a gut PPAR-a-striatal D1R pathway to reduce fat appetite in obese rats. Cell. Metab. 25, 335–344.

Liu, L., Zhang, S.X., Liao, W., Farhoodi, H.P., Wong, C.W., et al., 2017. Mechanoresponsive stem cells to target cancer metastases through biophysical cues. Sci. Transl. Med. 26, 9. eaan2966.

Zhang, C., Ni, D., Liu, Y., Yao, H., Bu, W., et al., 2017. Magnesium silicide nanoparticles as a deoxygenation agent for cancer starvation therapy. Nat. Nanotechnol. 12, 378–386.

CHAPTER FIFTEEN

Drugs and Drug-Related Topics

Contents

Biochemistry for Materials Science
https://doi.org/10.1016/B978-0-12-817054-0.00015-1
© 2019 Elsevier Inc.
All rights reserved.

15.1 PRODUCTION OF MALARIA DRUG BY DOUBLY ENGINEERED PLANTS

Artemisinin (see Fig. 15.1.1) is highly effective against drug-resistant malarial parasites (*Plasmodium falciparum*), which affects nearly half of the global population and kills >500,000 people each year. The cost of artemisinin is very expensive, because it is expensive to extract and purify the drug from *Artemisia annua*. Artemisinin was discovered by a Chinese scientist, Tu Youyou, who was awarded the Nobel Prize in 2015 for medicine. If the extraction and purification process is eliminated, this potent antimalarial drug will be available to the global population living in endemic regions.

Malhotra et al. (2016) developed the oral delivery of a nonprotein drug artemisinin biosynthesized ($\sim 0.8\,mg\,g^{-1}$ dry weight) at clinically meaningful levels in tobacco by engineering two metabolic pathways targeted to three different cellular compartments (chloroplast, nucleus, and mitochondria). The doubly transgenic lines showed a three-fold enhancement of isopentenyl pyrophosphate (IPP), which is a precursor of artemisinin, and the doubly modified pathways on chloroplasts resulted in higher expression and an efficient photo-oxidation of dihydroartemisinic acid to artemisinin.

Partially purified extracts from the leaves of transgenic tobacco plants inhibited in vitro growth progression of *P. falciparum*-infected red blood cells. Oral feeding of whole intact plant cells containing artemisinin reduced the parasitemia levels in the mice in contrast to the commercial drug. Such new approaches should facilitate low-cost production and delivery of

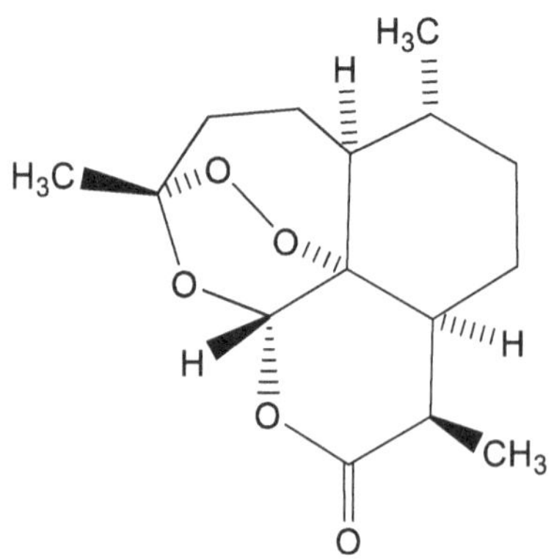

Fig. 15.1.1 Artemisinin.

artemisinin. This metabolic engineering method should be able to be applied to other drugs through other edible plants.

15.2 SPIDER VENOM PEPTIDE PROTECTS THE BRAIN AFTER STROKE

Stroke is the second most common cause of death worldwide. Six million people die from it each year, and 5 million survive but are left disabled. However, no drugs are available to protect the brain from stroke-related neuronal injury, which is caused by lack of oxygen. Acid-sensing ion channel 1a (ASIC1a), which is the primary acid sensor in mammalian (rodents') brains, detects a drop of brain pH, and reduces acidosis-induced neuronal death following cerebral ischemic stroke.

Chassagnon et al. (2017) found that Hi1a, a disulfide-rich spider venom peptide from the Australian funnel-web spider *Hadronyche infensa*, works as an inhibitor of ASIC1a, resulting in highly neuroprotective after ischemic stroke. Nuclear magnetic resonance (NMR) revealed that Hi1a is composed of two similar inhibitor cysteine knot (ICK) domains; these are separated by a short, structurally well-defined linker (see Fig. 15.2.1; Pineda et al., 2017).

Compared to known ASIC1a inhibitors, Hi1a induces a delay in the activation of ASIC1a. Hi1a binds to and stabilizes the closed state of the ASIC1a; thus it impedes the transition into a conducting state.

Intracerebroventricular (ICV) administration to rats of a single small dose of Hi1a (2 ng/kg) up to 8 h after stroke induction by occlusion of the middle

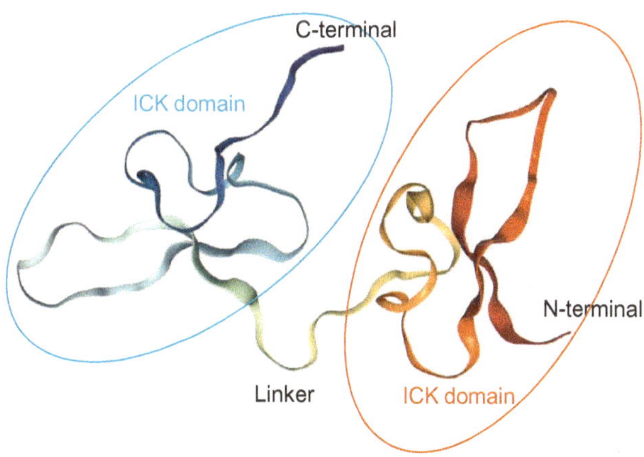

Fig. 15.2.1 Spider toxin-pi-hexatoxin-Hi1a as 2N8F in RCSB Protein Data Base (PDB). Two inhibitor cysteine knot (ICK) domains are connected by a short linker.

cerebral artery markedly reduced infarct size. In addition, ICV administration improved not only preservation of neuronal architecture but also neurological and motor functions.

In summary, Hi1a is a promising therapy to protect the brain from ischemic stroke. Furthermore, it could be a powerful tool for probing the role of ASIC1a in acid-related neuronal injury and various neurological disorders.

15.3 REDUCING BODY FAT BY ANTIBODY IN MICE

Menopause is associated with bone loss and enhanced visceral adiposity (increasing fat around internal organs). During menopause, levels of follicle-stimulating hormone (FSH) increase drastically. A polyclonal antibody that targets the β-subunit of the pituitary hormone follicle-stimulating hormone (FSH) increases bone mass in mice.

The polyclonal antibody raised against a 13-amino acid sequence (LVYKDPARPNTQK) of the β-subunit of FSH (FSHβ) inhibits ovariectomy-induced bone loss in mice.

Liu et al. (2017) have found that the blocking FSH by the antibody in mice also increases cellular mitochondrial density, activates brown adipose tissue, and enhances thermogenesis.

These actions result from the specific binding of the antibody to the β-subunit of FSH to block its action. The studies of Liu et al. (2017) uncover opportunities for treating obesity and osteoporosis simultaneously. However, this requires caution and more studies on whether the β-subunit of FSH applies to humans.

15.4 LICORICE COMPOUNDS CAN INHIBIT CYTOCHROME P450 ENZYME

Li et al. (2017) evaluated the interaction between licorice supplements and drug metabolism. Extracts of three botanically identified licorice species (*Glycyrrhiza glabra L.*, *Glycyrrhiza uralensis* Fish. ex DC., and *Glycyrrhiza inflata* Batalin) and 14 isolated licorice compounds were tested regarding inhibition of 9 cytochrome P450 enzymes, which are responsible for degradation of toxic molecules and pharmaceuticals in the body.

G. glabra showed moderate inhibitory effects against CYP2B6, CYP2C8, C9, and C19, and weak inhibition against CYP3A4 (testosterone). In contrast, *G. uralensis* strongly inhibited CYP2B6 and moderately inhibited CYP2C8, C9, and C19, and *G. inflata* strongly inhibited CYP2C and moderately inhibited CYP1A2, CYP2B6, CYP2D6, and CYP3A4 (midazolam).

The licorice compounds isoliquiritigenin, licoricidin, licochalcone A, 18β-glycyrrhetinic acid, and glycycoumarin (Figs.15.4.1A–E, respectively) inhibited one or more members of the CYP2C family of enzymes. Glycycoumarin and licochalcone A inhibited CYP1A2, but only glycycoumarin inhibited CYP2B6. Isoliquiritigenin, glabridin (Fig. 15.4.1F), and licoricidin inhibited CYP3A4, while licochalcone A (specific to *G. inflata* roots) was a mechanism-based inhibitor. The three licorice species commonly used in botanical dietary supplements have varying potential for drug-botanical interactions as inhibitors of cytochrome P450 families.

Thus, each species of licorice displays a unique profile of constituents with potential for drug interactions.

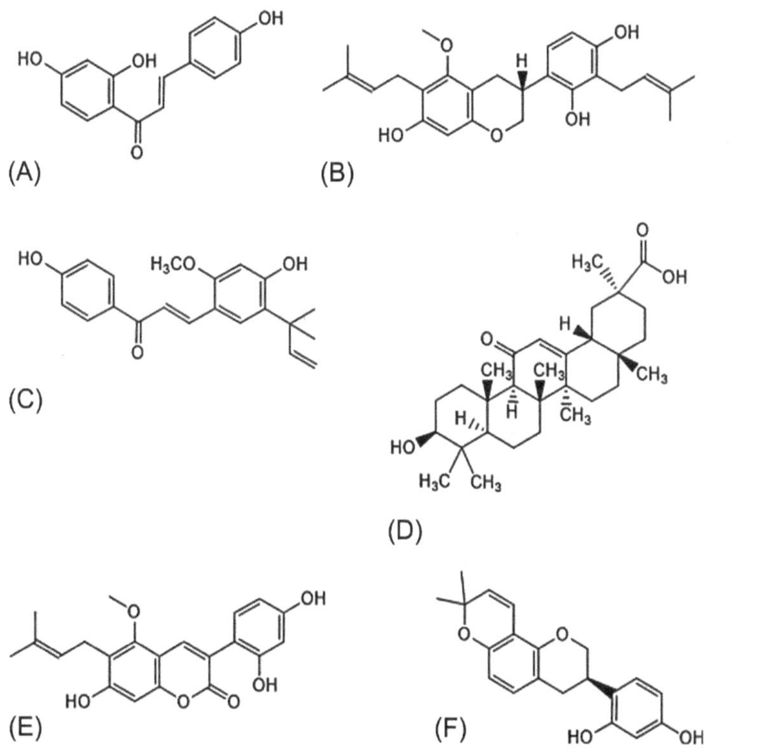

Fig. 15.4.1 (A) Isoliquiritigenin. (B) Licoricidin. (C). Licochalcone A. (D) 18β-Glycyrrhetinic acid. (E) Glycycoumarin. (F). Glabridin.

15.5 BRYOSTATIN 1 FOR HIV VIRUS ERADICATION

Scientists once accumulated 14 tons of the sea creature *Bugula neritina* to extract 18g of bryostatin 1 (see Fig. 15.5.1A). The macrocyclic organic compound is under study for its potential in HIV/AIDS eradication, cancer immunotherapy, and the treatment of Alzheimer's disease. However, as its availability is limited, research and clinical studies have been difficult.

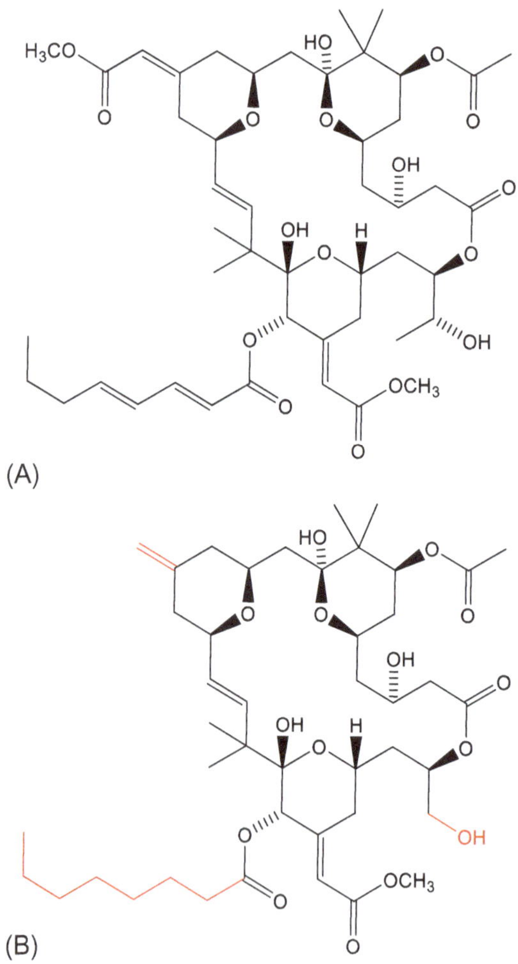

Fig. 15.5.1 (A) Bryostatin 1. (B) SUW133.

Wender et al. (2017) developed a 29-step synthesis of bryostatin 1 with 5% overall yield, and the synthesis of ~20 g per year became possible. Intermediates along the pathway can be modified straightforwardly to produce analogs.

For example, Marsden et al. (2017) synthesized a bryostatin analog, SUW133 (see Fig. 15.5.1B) and tested it in vivo.

15.6 ASTHMA DRUGS REDUCE PARKINSON'S DISEASE RISK

Drugs such as salbutamol (see Fig. 15.6.1A) can help a patient suffering an asthma attack to breathe. The drugs affect a protein receptor, which triggers the opening of the airways of lungs. Mittal et al. (2017) have found that drugs targeting this receptor (β2-adrenoreceptor; β2AR) may have another benefit in terms of reducing the risk of Parkinson's disease.

Parkinson's disease (PD) is related to the excess production of the protein α-synuclein; an accumulation of this protein forms characteristic clumps known as Lewy bodies. Mittal et al. (2017) found that β2AR is a regulator of the α-synuclein gene (SNCA). They found the β2AR agonist, such as salbutamol, a brain-penetrant asthma medication, reduced the risk of PD in a study spanning 11 years. In contrast, a β2AR antagonist, such as propranolol (see Fig. 15.6.1B), increased the risk.

15.7 HOW DOES SALINOMYCIN KILL CANCER STEM CELLS?

Cancer stem cells (CSCs) are cells within tumors that have a self-renewal ability and seeding capability of tumors. CSCs are very strong against normal treatments and related to metastasis and relapse. Salinomycin (see Fig. 15.7.1A) is one of the few agents against CSCs; however, the mechanism of how salinomycin works has not yet been solved.

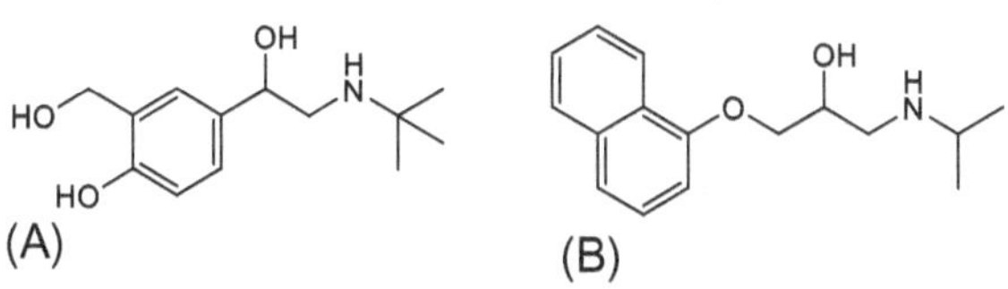

Fig. 15.6.1 (A) Salbutamol. (B) Propranolol.

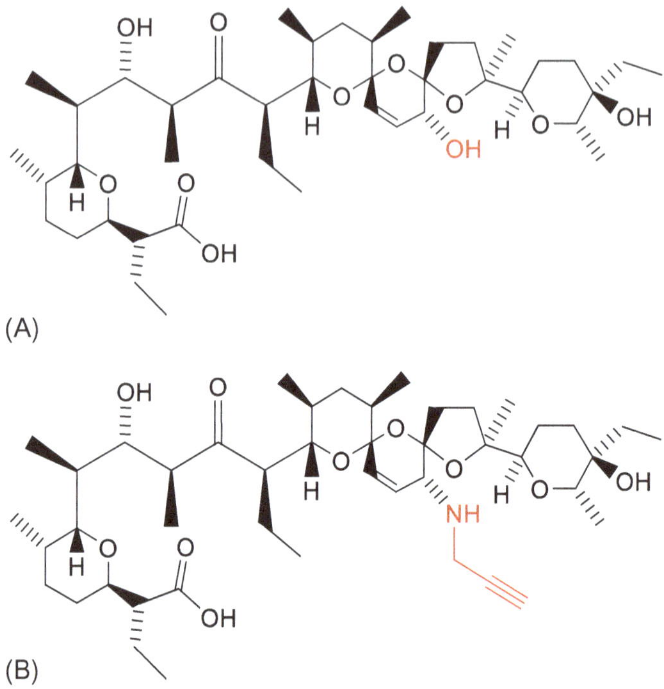

Fig. 15.7.1 (A) Salinomycin. (B) Ironomycin.

Mai et al. (2017) synthesized ironomycin (see Fig. 15.7.1B), which worked more effectively and selectively than salinomycin against breast CSCs in vitro and in vivo. Against their expectations, ironomycin gathered in lysosomes, and sequestered iron into lysosomes (see Figs.15.7.2A and B).

According to the depletion of cytoplasmic iron, cells triggered the degradation of ferritin (iron storing protein) in lysosomes. The so-called Fenton reaction, in which a high concentration of iron produces a reactive oxygen species, broke lysosome membranes, resulting in the death of CSCs (Fig. 15.7.2C). Alternatively, the low iron content makes CSCs more susceptible to ironomycin or salinomycin. This is the first study to reveal the mechanism of salinomycin.

15.8 HOW DO PLATINUM-CONTAINING DRUGS WORK?

15.8.1 Introduction

Cisplatin (see Fig. 15.8.1A) began to be used in clinics >40 years ago, despite a lack of understanding of the underlying cellular and molecular mechanisms

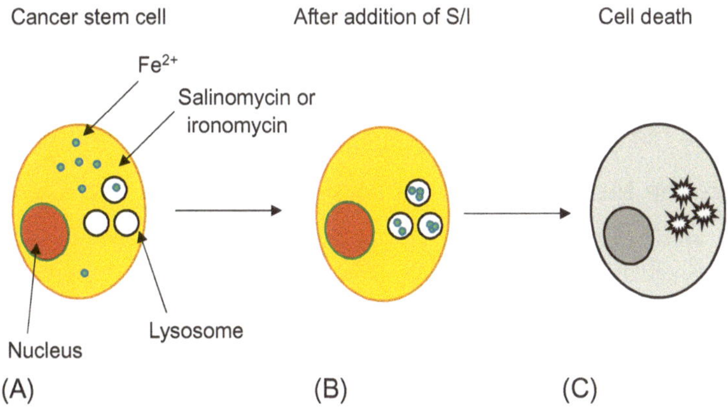

Fig. 15.7.2 A schematic diagram showing how the mechanism of ironomycin or salinomycin kills cancer stem cells. (A) In the cancer stem cell, ferrous iron ions exist ubiquitously. Salinomycin/ironomycin (S/I) are added to the cell. (B) Salinomycin/ironomycin gather into lysosomes together with ferrous iron. (C) Lysosomes then break because of too much iron in them or the cell is easily damaged because of low abundance of ferrous ions in the cytoplasma. Finally, the cell dies.

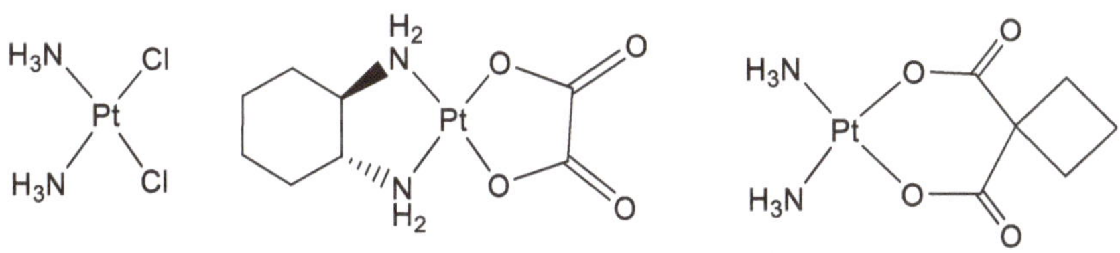

Fig. 15.8.1 Platinum-containing cancer drugs: (A) cisplatin. (B) oxaliplatin. (C) carboplatin.

(Kelland, 2007). However, cisplatin has been a component of regimens for at least 18 tumor types, such as germ cell, breast, and lung malignancies (Chu and DeVita, 2008).

To overcome the side effects of cisplatin and the resistance to it, two derivatives called oxaliplatin and carboplatin (see Figs. 15.8.1B and C, respectively) were developed; these have also been used clinically for a wide range of cancers.

It is interesting that oxaliplatin has different side-effect profiles from those of cisplatin and carboplatin. It is also used against colorectal and other gastrointestinal cancers, against which cisplatin and carboplatin have minimal efficacy. The choice to use oxaliplatin against colorectal cancer is made due to its activity against colorectal cancer cell lines without any rationale involving its mechanism. Although oxaliplatin is assumed to kill cancer cells, as cisplatin does, by eliciting a DNA damage response, no satisfactory reason for the unique use and side-effect profile of oxaliplatin has been found.

15.8.2 The Mechanisms of Action of Oxaliplatin and Cisplatin

Bruno et al. (2017) have demonstrated the fundamental mechanisms of oxaliplatin relative to those of cisplatin. They have also proposed which agents should be used following the mechanism for the treatment of cancer.

Bruno et al. (2017) show that oxaliplatin, unlike cisplatin and carboplatin, does not kill cells through the DNA-damage response. Instead, oxaliplatin kills cells by inducing ribosome biogenesis stress. This difference in drug mechanism explains the distinct clinical implementation of oxaliplatin relative to cisplatin, and will enable selection of distinct platinum-containing drugs for distinct malignancies based on the mechanism.

These data emphasize the functional difference of core components of up-to-date cancer therapy and the potential benefits of applying a mechanism-based dose of present and future anticancer drugs.

15.9 A NEW SONIC HEDGEHOG INHIBITOR

15.9.1 Hedgehog Signaling Pathway

The hedgehog (Hh) signaling pathway plays a central role in matters such as embryonic development, controlling cell growth, differentiation, and tissue patterning. Stimulation of the hedgehog signaling pathway is achieved by a complex sequence of molecular events at the level of the membrane and primary cilia of vertebrate cells, resulting in an intracellular signaling cascade and transcriptional response.

Canonical activation of this pathway is initiated by binding of the hedgehog signaling proteins (sonic (Shh), Indian (Ihh), and/or desert (Dhh) hedgehog) to the extracellular domain of the transmembrane receptor patched. This event relieves patched-mediated inhibition on the smoothened (Smo) receptor. Then Smo is allowed to translocate from the plasma membrane and endoplasmic vesicles to the primary cilium. Smo activation results in the accumulation of the active forms of Gli2 and Gli3 transcription factors, which stimulate the transcription of Gli-controlled genes, including Gli1 and PTCH1 (see Fig. 15.9.1).

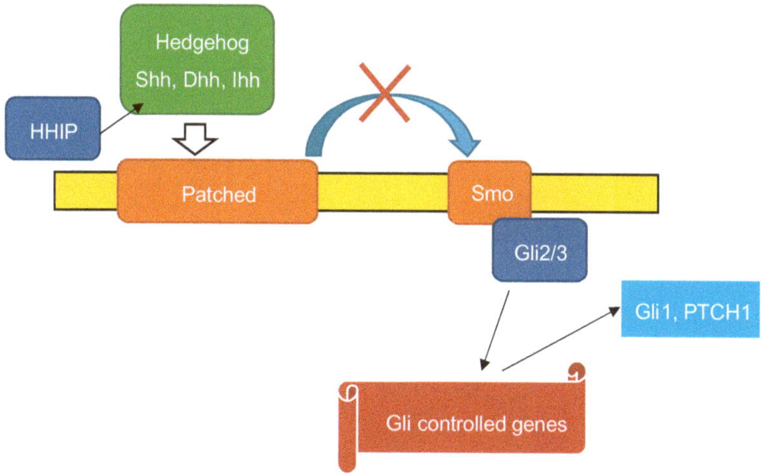

Fig. 15.9.1 Hedgehog signaling pathway. See text for explanation.

15.9.2 A new Sonic Hedgehog (Shh) Inhibitor

Aberrant activation of the Hh signaling pathway has relation to some human cancers. Major efforts toward the identification of chemical modulators of the hedgehog pathway have yielded several inhibitors, such as FDA-approved cyclopamine and vismodegib (see Fig. 15.9.2A).

In contrast, potent inhibitors of the sonic hedgehog/patched interaction, the most upstream event in ligand-induced activation of this signaling pathway, have been enigmatic. To reduce this gap, a genetically encoded cyclic peptide was designed based on the sonic hedgehog (Shh)-binding loop of hedgehog interacting protein (HHIP) and subjected to the screening of macrocyclic peptide libraries produced in *Escherichia coli* cells.

Thus Owens et al. (2017) obtained an optimized macrocyclic peptide inhibitor (HL2-m5; see Fig. 15.9.2B) that binds Shh with a 120-fold affinity compared to that of the parent molecule. HL2-m5 is able effectively to suppress Shh-mediated hedgehog signaling and Gli-controlled gene transcription in living cells. HL2-m5 is the most potent inhibitor of the sonic hedgehog/patched interaction reported so far.

This new macrocyclic peptide modulator of the hedgehog pathway is expected to be an important probe for investigating and targeting ligand-dependent hedgehog pathway activation in cancer and other pathologies. This work of Owens et al. (2017) also introduces a general strategy for the development of cyclopeptide inhibitors of protein-protein interactions.

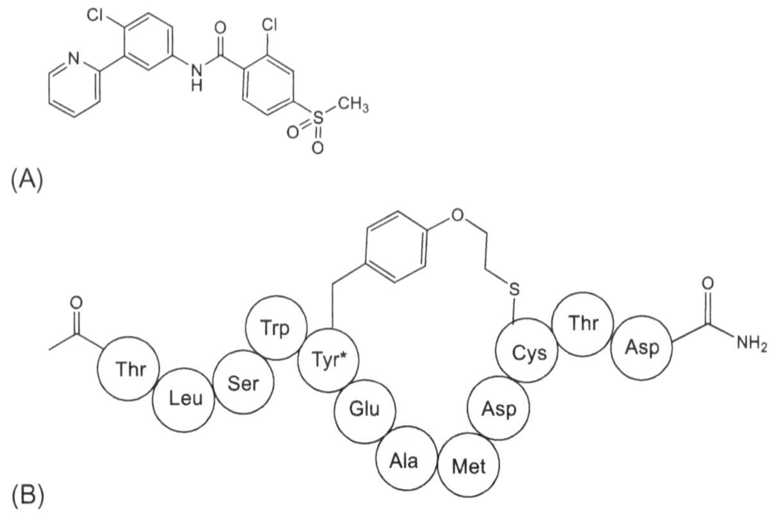

Fig. 15.9.2 (A) Vismodegib. (B) A new sonic hedgehog (Shh) inhibitor, HL2-m5. Tyr* is a tyrosine-like nonnatural amino acid which forms bridge with sulfur of cysteine. See Table 2.1.1 for abbreviation of the amino acids.

15.10 LASMIDITAN

15.10.1 Introduction

At the time of writing, Eli Lilly & Co. intends to buy CoLucid Pharmaceuticals, a biotech firm that is developing lasmiditan (see Fig. 15.10.1), an oral 5-HT_{1F} agonist (see Table 15.10.1) for the treatment of migraines. Lasmiditan has completed the first of two Phase III clinical trials. If the data are positive, lasmiditan could be approved by the FDA in 2018.

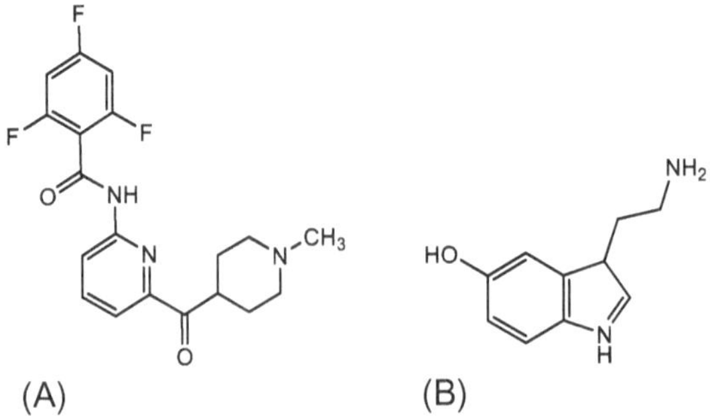

Fig. 15.10.1 (A) Lasmiditan. (B) Serotonin.

Table 15.10.1 Serotonin Receptor Agonists

Nonselective agonists
5-HT_1 receptor agonists
5-HT_{1A}, 5-HT_{1B}, 5-HT_{1C}, 5-HT_{1C}, 5-HT_{1D}, 5-HT_{1E}, 5-HT_{1F}
5-HT_2 receptor agonists
5-HT_{2A}, 5-HT_{2B}, 5-HT_{2C}
5-HT_3 receptor agonists
5-HT_4 receptor agonists
5-HT_{5A} receptor agonists
5-HT_6 receptor agonists
5-HT_7 receptor agonists

Eli Lilly & Co. discovered lasmiditan and licensed it to CoLucid in 2005, when the former company was not focused on pain management. Lasmiditan could be the "first significant innovation for the acute treatment of migraine in >20 years."

15.10.2 Serotonin Receptor Agonists

An agonist is a chemical that binds to a receptor and activates the receptor to produce a biological response. A serotonin receptor agonist is an agonist of one or more serotonin receptors. Serotonin is 5-hydroxytryptamine (5-HT), a neurotransmitter and hormone. Serotonin receptor agonists are classified as shown in Table 15.10.1.

15.11 CASPASES ARE NEW TARGETS OF NON-STEROIDAL ANTI-INFLAMMATORY DRUGS (NSAIDs)

15.11.1 Introduction to NSAIDs

Nonsteroidal antiinflammatory drugs (NSAIDs) are the most widespread pharmaceuticals in the world. For example, 30 billion doses are taken annually in the United States. The antiinflammatory properties of NSAIDs are used in analgesic and antifever treatments ranging from mild pain and rheumatoid arthritis to cancer.

The best-characterized antiinflammatory properties of NSAIDs come from inhibition of cyclooxygenase (COX) isozymes. COX inhibition reduces the production of prostaglandins, which are lipid autacoids related to various cellular processes such as angiogenesis, apoptosis, and cell migration. However, the existence and contributions of other NSAID targets are unclear.

These targets may relate to NSAID antiinflammatory mechanisms of reverse drug reactions. Reported NSAID functions include those such as activation of intrinsic and extrinsic pathways of apoptosis and activation of stress kinases. However, most of these effects appear only at super-physiological concentrations, limiting their biological relevance.

15.11.2 Introduction to Caspases

Cysteine-aspartic proteases (caspases) are a family of cysteine proteases that cleave peptide substrates where the aspartic acid exists. Caspases play essential roles in cell death, differentiation, and inflammation (Lamkanfi and Dixit, 2012).

Initiator caspases (caspase-2, -8, -9, and -10) contain large prodomains for recruitment into multiprotein complexes, such as inflammasomes and apoptosomes. Upon activation, initiator caspases will be proteolytically activated; subsequently they activate executioner caspases (caspase-3, -6, and -7; see Fig. 15.11.1A for caspase-3), resulting in cell death through apoptosis (Lamkanfi, 2011).

The inflammatory caspases (caspase-1, -4, and -5) generate biologically active cytokines such as IL-1b and IL-18, and induce cell death (e.g., Kayagaki et al., 2011). Due to the important role of caspases in inflammation and disease, it is essential to understand the contributions of caspase inhibition in NSAID pharmacology.

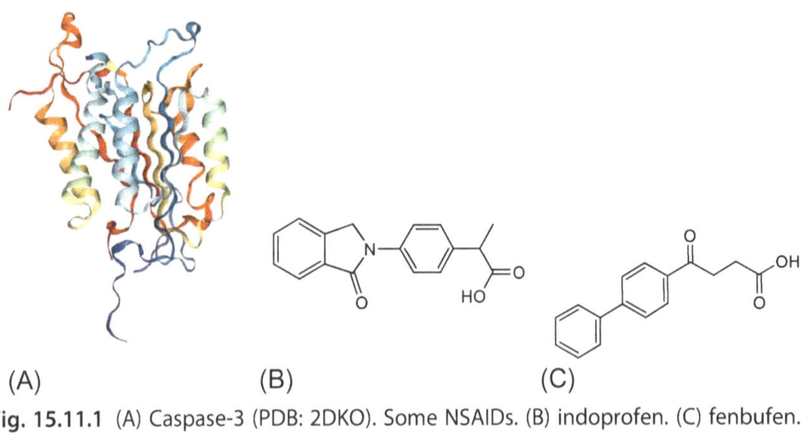

(A) (B) (C)

Fig. 15.11.1 (A) Caspase-3 (PDB: 2DKO). Some NSAIDs. (B) indoprofen. (C) fenbufen.

15.11.3 Caspases are New Targets of NSAIDs

Smith et al. (2017) have identified as novel targets for some NSAIDs (such as indoprofen (Fig. 15.11.1B), fenbufen (Fig. 15.11.1C), ibuprofen, ketorolac, etc.) at physiologic concentrations both in vitro and in vivo. They characterize caspase activity both in vitro and in cell culture; they also combine computational modeling and biophysical analysis to determine the mechanism of action. Smith et al. observe that inhibition of caspase catalysis reduces cell death and the generation of pro-inflammatory cytokines.

Furthermore, NSAID inhibition of caspases is COX independent, representing a new antiinflammatory mechanism. This finding expands upon existing NSAID antiinflammatory behaviors, with implications for patient safety and next-generation drug design.

15.12 MOLECULAR PROSTHETICS

15.12.1 Introduction to Molecular Prosthetics

Many metal ions play important roles in a wide variety of biological functions. To use such metals properly in cells, the transporting mechanisms of the metals must work without problems. Otherwise the metal accumulates or lacks in the cell.

Iron is one such important metal that is required in cells. Human diseases such as anemia, cystic fibrosis, arrhythmia, etc. are the result of deficiencies in iron-transporting proteins. Fig. 15.12.1A shows a schematic diagram of iron transportation. Through the cell membranes, which correspond to the outer cell membrane or the nucleus membrane, there are transmembrane proteins for iron, which are shown in Fig. 15.12.1A. Iron ions, or weakly bound iron ions on small organic molecules such as citrate, are transported through the protein by concentration gradients. For example, the transmembrane protein named DMT1 transports iron into the cell. Therefore, deficiencies of DMT1 cause deficiencies of iron in the cell. Mitoferrin (MFRN1) (and DMT1) transports iron into the mitochondria matrix through the mitochondrial membrane. Ferroportin (FPN1) is a transmembrane protein that transports iron from the inside to the outside of the cell.

15.12.2 Molecular Prosthetics

Grillo et al. (2017) tried to find a small lipophilic molecule that can transport iron ion through the cell membrane. They added the natural product

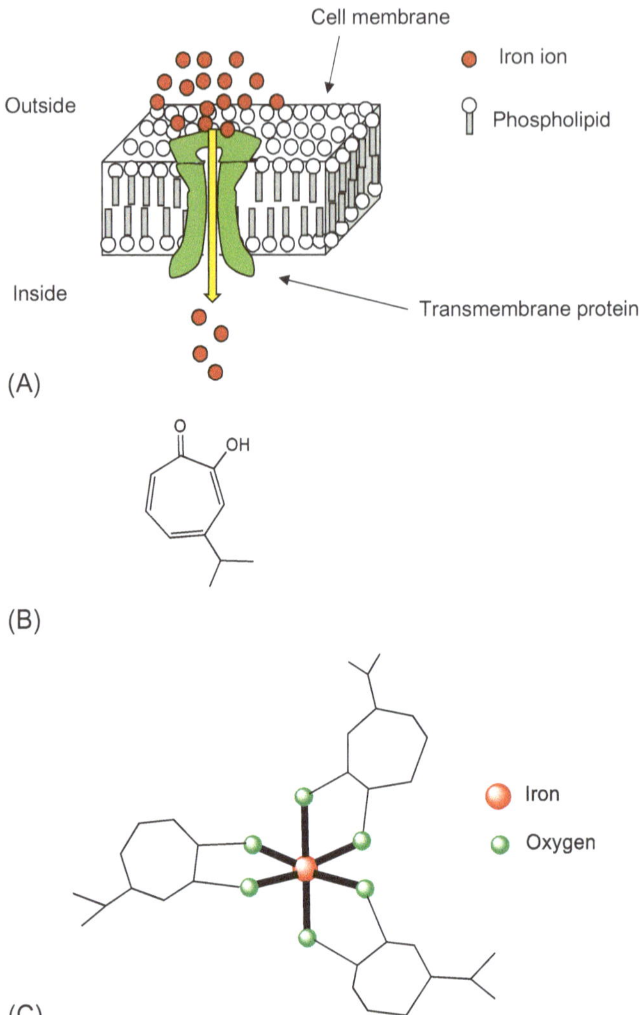

Fig. 15.12.1 (A) Schematic diagram of iron transportation. Cell membrane with a transmembrane protein is shown. Iron is transported through the protein by concentration gradients. (B) Hinokitiol. (C) Three molecules of hinokitiol bind to an iron ion.

hinokitiol (see Fig. 15.12.1B), which was isolated by Nozoe (1936) from the *Chamaecyparis taiwanensis* (Taiwan Hinoki) tree.

They cultured the yeast that lacks the iron transporting complex (Fet3Ftr1). The yeast cannot grow under iron-limiting conditions. However, they found that the yeast can survive under the iron-conditions with hinokitiol. As shown in Fig. 15.12.1C, both ferric and ferrous iron ions form a complex with three hinokitiol molecules.

Furthermore, Grillo et al. engineered zebrafish with lacking the mitoferrin. The zebrafish showed lower levels of hemoglobin. However, when hinokitiol was added, the hemoglobin levels returned to normal.

These experiments using the small molecule to the cell and the animal were successful as molecular prosthetics. Although they cannot be applied directly to human diseases, Grillo et al.'s findings have large scientific and therapeutic implications.

15.12.3 New Roles of Antibiotics of Dithiolopyrrolones (DTPs)

Natural products have unique and complex structures that provide valuable antibiotic functions. As antibiotic resistance increases, natural products are expected to have new antimicrobial functions, because their functions are not understood perfectly.

Dithiolopyrrolones (DTPs) are disulfide-containing natural products that are not yet well explored; examples are holomycin and thiolutin, which are shown in Figs.15.12.2A and B, respectively. They show potential antimicrobial activities against multidrug-resistant pathogens, such as *Staphylococcus aureus* (MRSA). DTPs were thought to target RNA polymerase, but conflicting results were also observed.

Using a chemical genomics screen in *E. coli*, Chan et al. (2017) uncovered a mode of action for DTPs: the disruption of metal homeostasis.

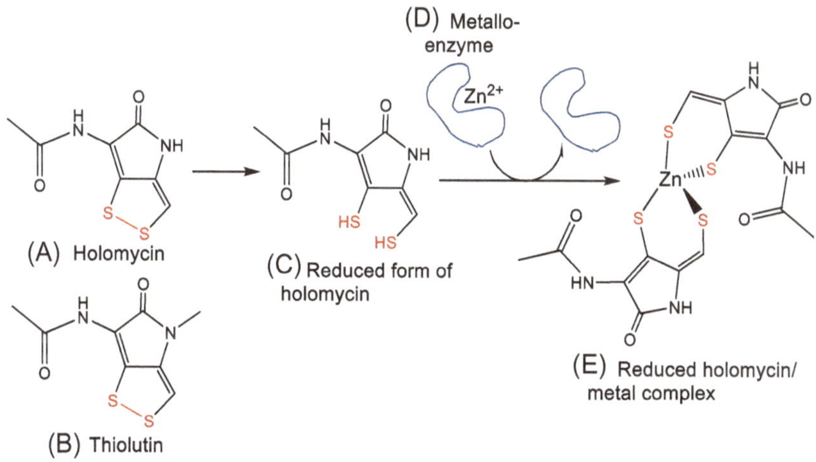

Fig. 15.12.2 Mechanism of holomycin's action. Reduced form of holomycin form a chelate with metal ion such as Zn^{2+} in metalloenzymes and removes it from metalloenzymes. (A) Holomycin. (B) Thiolutin. (C) Reduced form of holomycin. (D) Metalloenzyme containing Zn^{2+}. (E) Reduced holomycin/metal complex.

They showed that holomycin, a prototypical DTP, is reduced intracellularly (Fig. 15.12.2C), and activated and reduced holomycin chelates zinc with high affinity (Fig. 15.12.2D and E).

Examination of reduced holomycin against zinc-dependent metalloenzymes revealed that it inhibits *E. coli* class II fructose bisphosphate aldolase, but not RNA polymerase. Reduced holomycin also strongly inhibits metallo–β-lactamases (MBL; major contributors to clinical carbapenem resistance) in vitro, by removing the active site zinc.

These results suggest that holomycin is an intracellular metal-chelating antibiotic that inhibits a metalloenzymes, and that RNA polymerase is unlikely to be the primary target. This study indicates a link between the chemical structures of DTPs and their antimicrobial action. This study also emphasizes the potential of metal-chelating natural products in antimicrobial therapy.

15.13 SEARCH FOR THE CASE THAT TWO DRUGS ARE BETTER THAN ONE

Drug combinations often result in deleterious effects, but Licciardello et al. (2017) established a new method for searching the case that two drugs are better than one.

More than 30,000 drugs have been approved by the U.S. Food & Drug Administration. This is too many to search; therefore, Licciardello et al. (2017) eliminated redundant drugs, and chose 954 active small molecules. They further grouped the compounds into classes based on their structures and known activities, and then used software to help with their selection. They finally made a list of 308 compounds; this is called the CeMM Library of Unique Drugs (CLOUD).

They screened pairwise combinations of CLOUD drugs for damaging cancer cells, and discovered a synergistic interaction between a prostate cancer drug flutamide (see Fig. 15.13.1A) and an antithrombotic agent

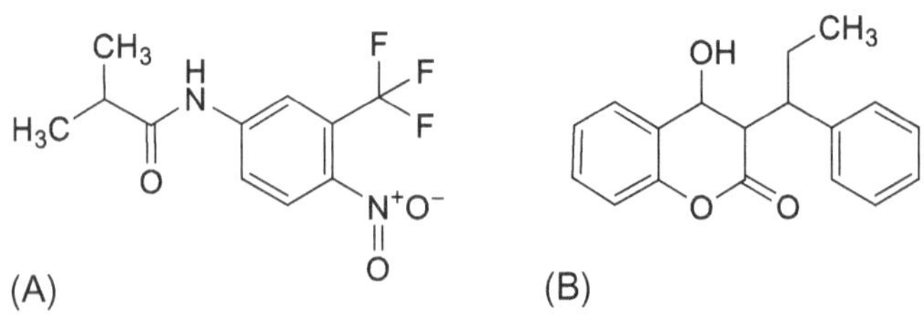

Fig. 15.13.1 (A) Flutamide. (B) Phenprocoumon (PPC).

phenprocoumon (PPC; see Fig. 15.13.1B). The combination of these drugs was effective in killing a drug-resistant form of prostate cancer. Combinations of non-CLOUD drugs in the same classes as flutamide and phenprocoumon also fight drug-resistant cancer, helping to confirm that the CLOUD compounds successfully represent their classes and that their reduction policy was valid.

Thus, the CLOUD compounds are the ideal sets of drugs to discover the effective combination of two drugs.

15.14 NANOFIBER COATINGS USING QUORUM SENSING

15.14.1 Introduction to Antimicrobial Strategy

A wide range of materials, such as interventional devices, implanted objects, and dressings, can be contaminated by bacteria. One effective strategy to prevent these bacterial infections is to design materials and surface coatings that release agents that can prohibit bacterial cell proliferation or biofilm formation. These surface-intermediated approaches have practical advantages to the systemic administration of antimicrobial agents.

The use of conventional antimicrobial agents has increased antibiotic-resistant strains of many common human pathogens. For example, methicillin-resistant strains of the Gram-positive pathogen *S. aureus* (MRSA) kill more people in the United States annually than Parkinson's disease, HIV/AIDS, emphysema, and homicide together. In view of this rapidly emerging threat, new methods for preventing bacterial infections not by conventional antibiotics but by other biocidal strategies are urgently needed.

One potential alternative to these biocidal strategies is the use of agents that can inhibit bacterial "quorum sensing" (QS), which is a mechanism of bacteria and governs many infection phenotypes in common pathogens, including virulence factor production and biofilm formation, but that do not promote bacterial cell death. This "antivirulence" approach is promising in many clinically relevant bacteria, including *S. aureus* and the Gram-negative pathogens *Pseudomonas aeruginosa* and *Acinetobacter baumannii*.

15.14.2 Nanofiber Coatings That Inhibit Quorum Sensing (QSI)

Kratochvil et al. (2017) have reported that the fabrication and biological evaluation of nonwoven polymer nanofiber coatings that inhibit quorum sensing (QS) and virulence in the human pathogen *S. aureus*. Their results

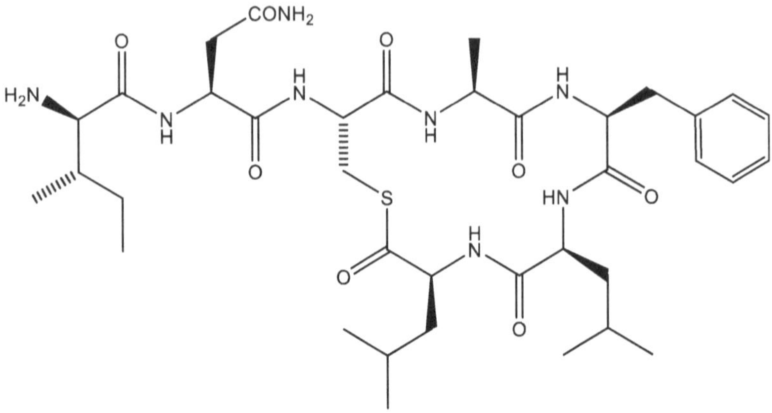

Fig. 15.14.1 The peptide inhibits quorum sensing in *S. aureus.*

show that macrocyclic peptide (see Fig. 15.14.1), a potent and synthetic nonbactericidal quorum sensing inhibitor (QSI) in *S. aureus* (Broderick et al., 2014), can be loaded into degradable polymer nanofibers by electrospinning, and that this approach can deposit QSI-loaded nanofiber coatings onto model nonwoven mesh substrates.

The QSI retained its biological activity over ~3 weeks, and strongly inhibited agr-based QS in a GFP reporter strain of *S. aureus* for at least 14 days without promoting cell death. These materials also inhibited production of hemolysins (proteins that cause rupture of red blood cell), a QS-controlled virulence phenotype, and reduced the lysis of erythrocytes (rupture of red blood cell) when placed in contact with wild-type *S. aureus* growing on surfaces.

This approach can be used with many different polymers, active agents, and processing parameters to fabricate nanofiber coatings on surfaces important in healthcare contexts. *S. aureus* is one of the most common causative agents of bacterial infections in humans, and strains of this pathogen have developed significant resistance to conventional antibiotics.

The QSI-based strategies are new antiinfective materials and novel treatment strategies that target virulence to battle with growth of *S. aureus*. This approach also provides a cell culture example that could prove useful in future studies on the influence of QS modulation on the development and structure of bacterial communities.

15.15 A NEW ANTIPLATELET DRUG WITH LOW BLEEDING RISK

Antiplatelet agents are effective to cardiovascular (CV) and cerebrovascular diseases. However, the existing drugs have risks of unwanted and life-threatening bleeding that limit drug use or dosage. An antiplatelet drug with strong efficacy and low bleeding risk is needed.

Thrombin is a potent platelet drug that directly induces platelet activation via the G protein (heterotrimeric guanine nucleotide-binding protein)-coupled **p**rotease-**a**ctivated **r**eceptors PAR1 and PAR4.

A PAR1 antagonist has been approved for clinical use, but this is limited by a substantial bleeding risk. Conversely, an antiplatelet target of PAR4 has not been well characterized. Using anti-PAR4 antibodies, Wong et al. (2017) demonstrated a low bleeding risk and an effective antithrombotic profile with PAR4 inhibition in guinea pigs.

Screening and an extensive medicinal chemistry effort of Wong et al. (2017) resulted in the discovery of BMS-986120 (see Fig. 15.15.1), an orally active, selective, and reversible PAR4 antagonist. In a cynomolgus monkey arterial thrombosis model, BMS-986120 demonstrated highly efficacious antithrombotic activity. BMS-986120 also exhibited a low bleeding liability and a markedly wider therapeutic window compared to the standard antiplatelet agent.

These preclinical findings define the biological role of PAR4 in mediating platelet aggregation. In addition, they indicate that targeting PAR4 is an attractive antiplatelet target with the potential to treat patients at a high risk of atherothrombosis with superior safety.

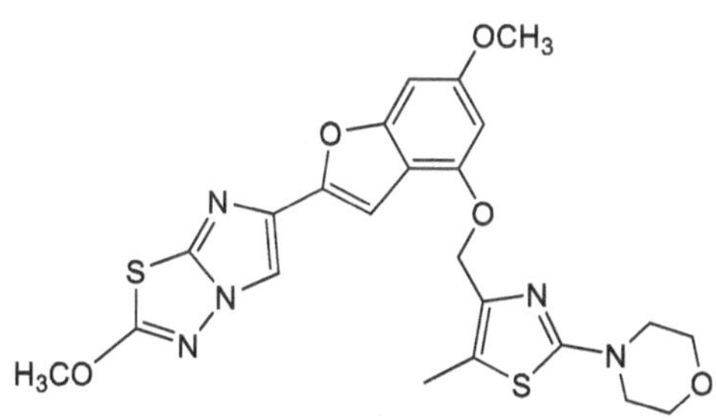

Fig. 15.15.1 BMS-986120.

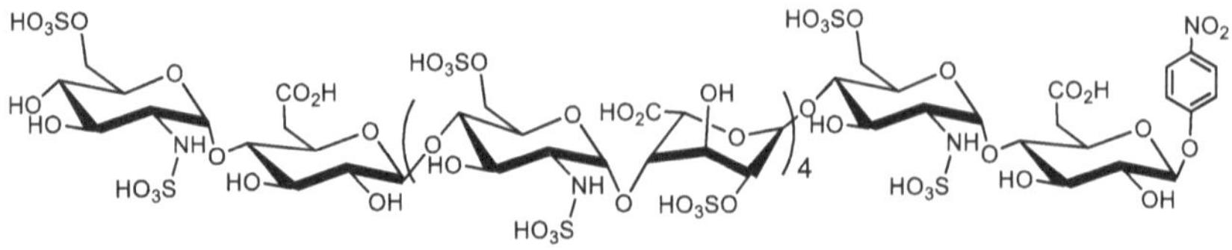

Fig. 15.16.1 12-mer-1.

15.16 NEW SYNTHETIC HEPARIN

Full-sized and low-molecular-weight heparins are widely used to treat a variety of clotting disorders. Although low-molecular-weight heparins (LMWH) are safer and more convenient to use than full-sized heparin, LMWH are still animal-derived products and have risks of contamination and supply chain interruptions that prevent standardization and reversibility of anticoagulation or blood-thinning.

Xu et al. (2017) demonstrated the synthesis of heparin dodecasaccharides (12-mers) at the gram scale. The anticoagulant activity of 12-mer-1 (see Fig. 15.16.1), which is one of these heparins, could be reversed by protamine ex vivo. A 7-day toxicity study of 12-mer-1 in a rat model showed no toxic effects. The data suggest that a synthetic homogeneous oligosaccharide can replace animal-sourced LMWHs.

REFERENCES

Broderick, A.H., Stacy, D.M., Tal-Gan, Y., Kratochvil, M.J., Blackwell, H.E., et al., 2014. Surface coatings that promote rapid release of peptide-based AgrC inhibitors for attenuation of quorum sensing in *Staphylococcus aureus*. Adv. Healthc. Mater. 3, 97–105.

Bruno, P.M., Liu, Y., Park, G.Y., Murai, J., Koch, C.E., et al., 2017. A subset of platinum-containing chemotherapeutic agents kills cells by inducing ribosome biogenesis stress. Nat. Med. 23, 461–471.

Chan, A.N., Shiver, A.L., Wever, W.J., Razvi, S.Z.A., Traxler, M.F., Li, B., 2017. Role for dithiolopyrrolones in disrupting bacterial metal homeostasis. PNAS 114, 2717–2722.

Chassagnon, I.R., McCarthy, C.A., Chin, Y.K.-Y., Pineda, S.S., Keramidas, A., et al., 2017. Potent neuroprotection after stroke afforded by a double-knot spider-venom peptide that inhibits acid-sensing ion channel 1a. PNAS 114, 3750–3755.

Chu, E., DeVita Jr., V.T., 2008. Physician's Cancer Chemotherapy Drug Manual. Jones & Bartlett Publishers, Sudbury, MA.

Grillo, A.S., Santa Mariia, A.M., Kafina, M.D., Cioffi, A.G., Huston, N.C., et al., 2017. Restored iron transport by a small molecule promotes absorption and hemoglobinization in animals. Science 356, 608–616.

Kayagaki, N., Warming, S., Lamkanfi, M., Vande Walle, L., Louie, S., et al., 2011. Non-canonical inflammasome activation targets caspase-11. Nature 479, 117–121.

Kelland, L., 2007. The resurgence of platinum-based cancer chemotherapy. Nat. Rev. Cancer 7, 573–584.

Kratochvil, M.J., Yang, T., Blackwell, H.E., Lynn, D.M., 2017. Nonwoven polymer nanofiber coatings that inhibit quorum sensing in *Staphylococcus aureus*: toward new non-bactericidal approaches to infection control. ACS Infect. Dis. 3, 271–280.
Lamkanfi, M., 2011. Emerging inflammasome effector mechanisms. Nat. Rev. Immunol. 11, 213–220.
Lamkanfi, M., Dixit, V.M., 2012. Inflammasomes and their roles in health and disease. Annu. Rev. Cell Dev. Biol. 28, 137–161.
Li, G., Simmler, C., Chen, L., Nikolic, D., Chen, S.-N., et al., 2017. Cytochrome P450 inhibition by three licorice species and fourteen licorice constituents. Eur. J. Pharm. Sci. 109, 182–190.
Licciardello, M.P., Ringler, A., Markt, P., Klepsch, F., Lardeau, C.-H., et al., 2017. A combinatorial screen of the CLOUD uncovers a synergy targeting the androgen receptor. Nat. Chem. Biol. 13, 771–778.
Liu, P., Ji, Y., Yuen, T., Rendina-Ruedy, E., DeMambro, V.E., et al., 2017. Blocking FSH induces thermogenic adipose tissue and reduces body fat. Nature 546, 107–112.
Mai, T.T., Hamaï, A., Hienzsch, A., Cañeque, T., Müller, S., et al., 2017. Salinomycin kills cancer stem cells by sequestering iron in lysosomes. Nat. Chem. 9, 1025–1033.
Malhotra, K., Subramaniyan, M., Rowat, K., Kalamuddin, M., Qureshi, M.I., et al., 2016. Compartmentalized metabolic engineering for artemisinin biosynthesis and malaria treatment by oral delivery of plant cells. Mol. Plant 9, 1464–1477.
Marsden, M.D., Loy, B.A., Wu, X., Ramirez, C.M., Schrier, A.J., et al., 2017. In vivo activation of latent HIV with a synthetic bryostatin analog effects both latent cell "kick" and "kill" in strategy for virus eradication. PLoS Pthog. 13, e1006575.
Mittal, S., Bjørnevik, K., Im, D.S., Flierl, A., Dong, X., et al., 2017. β2-Adrenoreceptor is a regulator of the α-synuclein gene driving risk of Parkinson's disease. Science 357, 891–898.
Nozoe, T., 1936. Uber die farbstoffe im holzteile des 'Hinoki' baumes. I. Hinokitin und hinokitiol. Bull. Chem. Soc. Jpn. 11, 295–298.
Owens, A.E., de Paola, I., Hansen, W.A., Liu, Y.-W., Khare, S.D., et al., 2017. Design and evolution of a macrocyclic peptide inhibitor of the sonic hedgehog/patched interaction. J. Am. Chem. Soc. 139, 12559–12568.
Pineda, S.S., Chin, Y.K.-Y., Senff, S., Mobli, M., Escoubas, P., et al., 2017. Potent neuroprotection after stroke afforded by a double-knot spider-venom peptide that inhibits acid-sensing ion channel 1a. Proc. Natl. Acad. Sci. U. S. A. 114 (14), 3750–3755.
Smith, C.E., Soti, S., Jones, T.A., Nakagawa, A., Xue, D., et al., 2017. Non-steroidal anti-inflammatory drugs are caspase inhibitors. Cell Chem. Biol. 24, 281–292.
Wender, P.A., Hardman, C.T., Ho, S., Jeffreys, M.S., Maclaren, J.K., et al., 2017. Scalable synthesis of bryostatin 1 and analogs, adjuvant leads against latent HIV. Science 358, 218–223.
Wong, P.C., Seiffert, D., Bird, E., Jeffrey, S., Watson, C.A., Bostwick, J.S., et al., 2017. Blockade of protease-activated receptor-4 (PAR4) provides robust antithrombotic activity with low bleeding. Sci. Transl. Med. 9, https://doi.org/10.1126/scitranslmed.aaf5294. eaaf5294.
Xu, Y., Chandarajoti, K., Zhang, X., Pagadala, V., Dou, W., et al., 2017. Synthetic oligosaccharides can replace animal-sourced low–molecular weight heparins. Sci. Transl. Med. 9, https://doi.org/10.1126/scitranslmed.aan5954. eaan5954.

FURTHER READING

Adams, B., 2017. Lilly buys migraine biotech CoLucid, and the drug it outlicensed, for $960M. FierceBiotech. Jan 18, 2017 7:18am.
Ganesan, R., Mittl, P.R.E., Jelakovic, S., Grutter, M.G., 2006. Extended substrate recognition in Caspase-3 revealed by high resolution X-ray structure analysis. J. Mol. Biol. 359, 1378–1388.

CHAPTER SIXTEEN

Detoxification of Toxic Gases of VX and Sarin

Contents

16.1 INTRODUCTION TO VX

In 2017, Jong-nam Kim was said to have been assassinated using *O*-ethyl *S*-[2-(diisopropylamino)ethyl] methylphosphonothioate (VX) in front of security cameras at the airport. He was half-brother to Jong-un Kim, the Chairman of the Workers' Party of Korea (WPK) and supreme leader of the Democratic People's Republic of Korea (DPRK).

This episode may remind Japanese people of when the Japanese cult Aum Shinrikyo killed one person and injured two more using VX in 1994–1995; they were the first victims of VX.

V-type nerve agents (structural formulas of VX-families are shown in Fig. 16.1.1) are toxic organophosphonates (OPs), and used as chemical weapons. The materials that have phosphono-thioester bonds with the amino group in the side chain are harmful, being VX.

VX is a tasteless and odorless liquid with an amber-like color. It destroys the nervous system of the body. As little as ten milligrams is fatal for a human through skin contact. Therefore, detoxifying VX in vitro (out of the body) as well as in vivo (in the body) is very important.

In 2018, Sergei Skripal, who was a retired Russian military intelligence officer, and his daughter, Yuliya were poisoned on March 4 in Salisbury, England. U.K. Prime Minister Theresa May said, "It is now clear that

Biochemistry for Materials Science
https://doi.org/10.1016/B978-0-12-817054-0.00016-3

© 2019 Elsevier Inc.
All rights reserved.

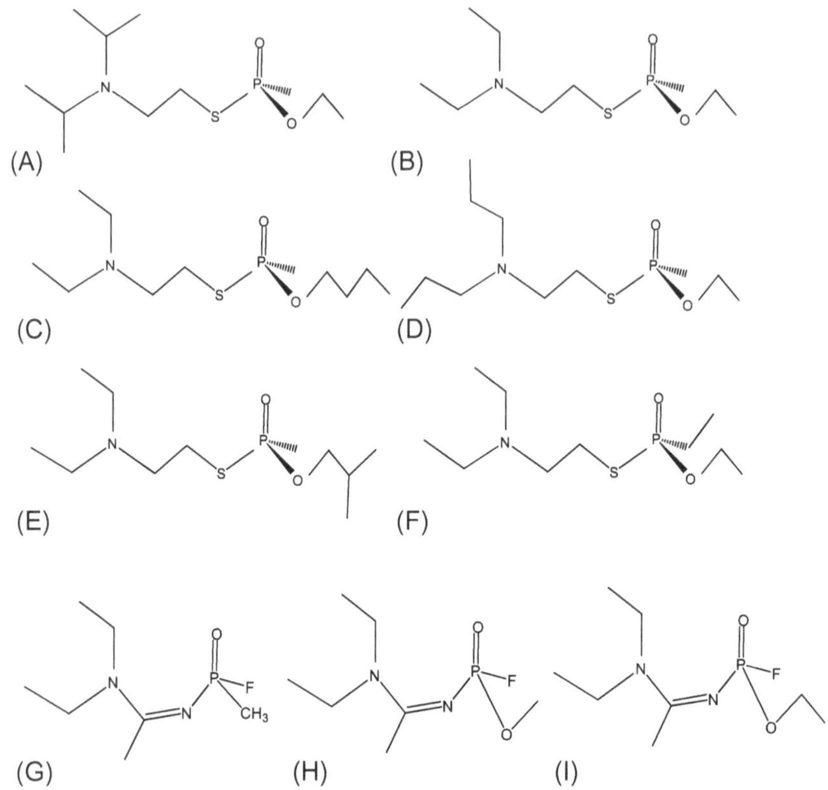

Fig. 16.1.1 Structural formulas of V-type nerve agents. (A) VX, (B) VM, (C) CVX, (D) n-propyl VX, (E) VR, (F) VE. A family of "Novichok" poisons. (G) A-230. (H) A-232. (I) A-234.

Mr. Skripal and his daughter were poisoned with a military-grade nerve agent of a type developed by Russia." VR-family nerve agents such as A-230, A-232, and A-234, which are called "Novichok" poisons (see Fig. 16.1.1G–I) could have been used (Peplow, 2018).

16.2 INTRODUCTION TO SARIN

Sarin is a colorless, odorless, and venomous liquid, which is used as a chemical weapon, because it is one of the strongest nerve agents. Sarin is an organophosphorus compound expressed as $[(CH_3)_2CHO]CH_3P(O)F$, of which the IUPAC name is Propan-2-yl-methylphosphonofluoridate (see Fig. 16.2.1).

Sarin has history regarding the politics of America. In 2003, the Iraq War started with the invasion of the US-led troops ordered by their President,

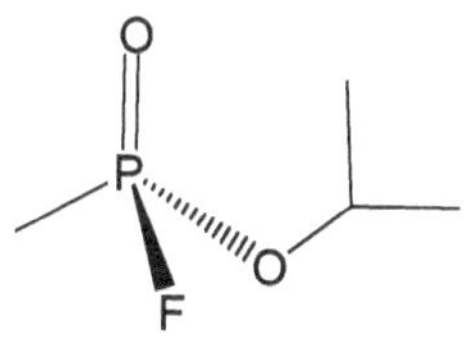

Fig. 16.2.1 Sarin.

George Bush. The reason for the invasion was that Iraq was believed to possess chemical weapons of mass destruction including sarin, and to use them in the military. In April 2017, the U.S. President, Donald Trump, ordered the strikes by cruise missiles on Syria due to believed suspicious use of sarin.

In 1995, the Japanese cult Aum Shinrikyo caused the Tokyo subway sarin attack, in which 12 persons were killed, 50 were severely injured, and about 5000 more were injured.

16.3 HOW VX WORKS AS TOXIC MATERIALS IN THE HUMAN BODY

How VX works is schematically shown in Fig. 16.3.1. This figure shows the nerve terminal when VX does not exist. The nerve signal is transmitted from the terminal of the presynaptic cell to the target cell. There are synaptic vesicules at the terminal of the cell, in which acetylcholine (ACh) is accumulated (shown as dots in Fig. 16.3.1A-1). When the nerve signal comes, the synaptic vesicules are opened, and ACh is released. ACh attaches at the ion channel of the target cell, and opens the ion channels. This allows Na^+ ions to enter the target cell (Fig. 16.3.1A-2). This changes the membrane potential of the target cell, resulting in transmitting signals from the terminal cell to the target cell.

After a few tens of milliseconds, acetylcholinesterase (AChE), which exists ubiquitously at the synapse, attaches to the ACh (Fig. 16.3.1A-3). Finally, both ACh and AChE are released from the ion channel, resulting in the ion channel closing (Fig. 16.3.1A-4). The used ACh is scavenged and returned into the terminal of the cell, and ACh and AChE are prepared to be reused.

When VX (diamonds) exists, VX reacts with acetylcholinesterase (AChE) (Fig. 16.3.1B-3′). In this form, AChE cannot react with ACh, which is attached to the ion channel. As result, the ion channel is kept open, and the signal cannot be transmitted properly at the synapse. Ultimately, this causes death.

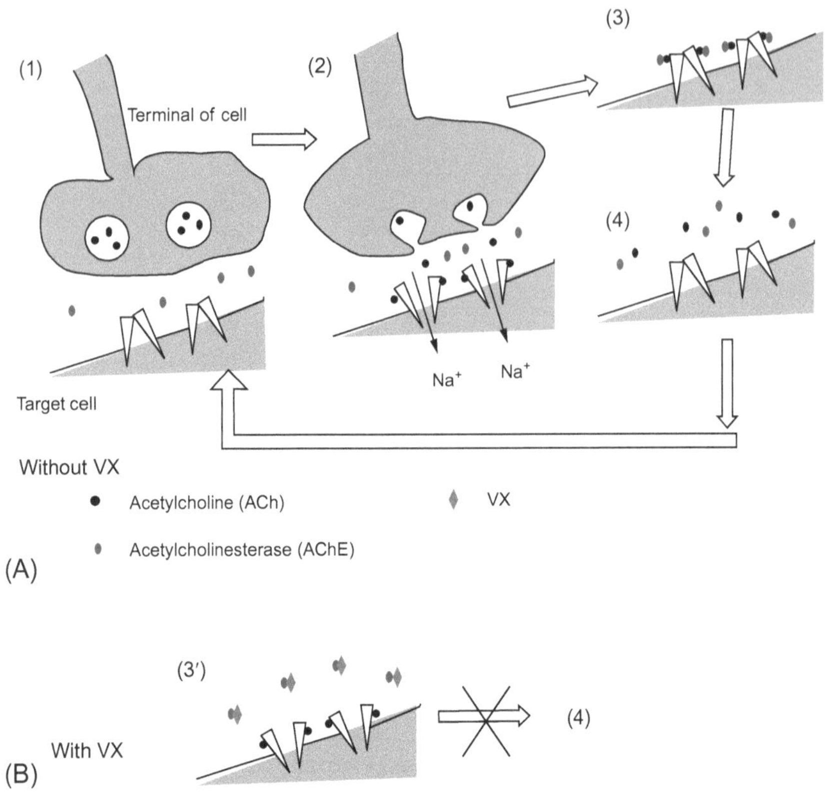

Fig. 16.3.1 Nerve terminal and function of acetylcholine (ACh). (A) Without VX and (B) with VX.

Sarin works similarly to VX, as it also reacts with AChE, so that ACh is not decomposed. Thus, the ion channel is kept open and the signal cannot be transmitted properly at the synapse any more.

16.4 DETOXIFICATION OF VX

Schneider et al. (2016) succeeded in detoxifying VX using sulfonatocalix[4]arenes which are strong cation binders in water. These calixarenes were decorated with a substituent containing a hydroxamic acid, which is known to detoxify organophosphonates (OPs) (e.g. Perret et al., 2006; Guo and Liu, 2014; Hof, 2016). The compounds shown in Fig. 16.4.1A (X=N) and Fig. 16.4.1B (X=CH) have highly promising

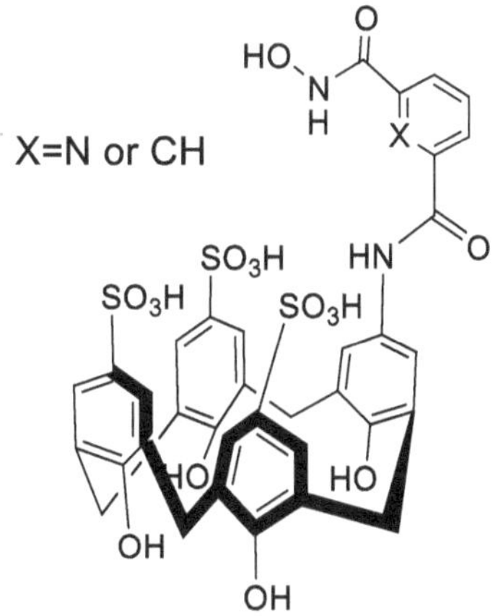

Fig. 16.4.1 Sulfonatocalix[4]arenes. (A) X=N. (B) X=CH.

properties under 37°C and pH7.40 (Schneider et al., 2016). Their study showed that both compounds effectively detoxified VX (Fig. 16.1.1A) as well as V-type nerve agents (Fig. 16.1.1B–F).

The compound of Fig. 16.4.1A detoxified 78% of VX within the first hour under 37°C and pH7.40 at the same amount, and 94% under two equivalent of VX. The compound of Fig. 16.4.1B detoxified similarly to that of Fig. 16.4.1A. However, the residual VX reached plateaus of 5% and 20% after 24h with one equivalent of the compounds of Fig. 16.4.1A and B, respectively. As a result, the calixarenes mediate the selective cleavage of the P—S bond of VX, and efficiently suppress the formation of the toxic metabolite (Schneider et al., 2016).

Sarin is vulnerable to heat and water (see Fig. 16.4.2). When sarin (1) is thrown into water, it is easily decomposed into hydrogen fluoride (HF) and isopropyl methylphosphonate (2) by hydrolysis. Isopropyl methylphosphonate (2) further becomes (3) methylphosphonic acid and (4) isopropyl alcohol.

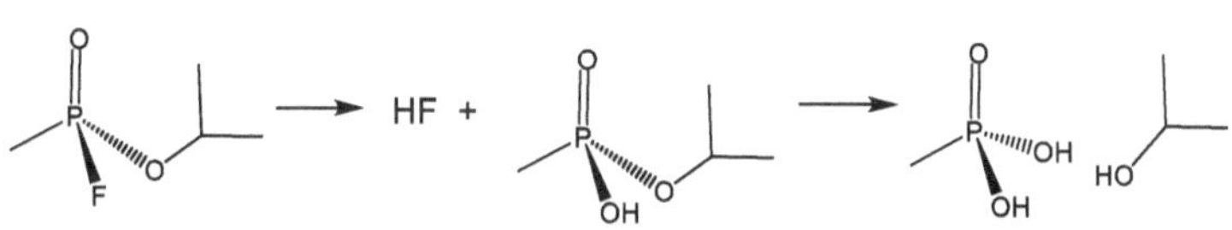

(A) 1 (B) 2 (C) 3 4

Fig. 16.4.2 Hydrolysis of sarin.

REFERENCES

Guo, D.-S., Liu, Y., 2014. Supramolecular chemistry of p-sulfonatocalix[n]arenes and its biological applications. Acc. Chem. Res. 47, 1925–1934.

Hof, F., 2016. Host-guest chemistry that directly targets lysine methylation: synthetic host molecules as alternatives to bio-reagents. Chem. Commun. 52, 10093–10108.

Peplow, M., 2018. Nerve agent attack on spy used 'Novichok' poison. C&EN 96 (12), 3.

Perret, F., Lazar, A.N., Coleman, A.W., 2006. Biochemistry of the para-sulfonato-calix[n] arenes. Chem. Commun. 0, 2425–2438.

Schneider, C., Bierwisch, A., Koller, M., Worek, F., Kubik, S., 2016. Detoxification of VX and other V-type nerve agents in water at 37°C and pH 7.4 by substituted sulfonatocalix[4]arenes. Angew. Chem. Int. Ed. 55, 12668–12672.

CHAPTER SEVENTEEN

Decomposing Explosives and Reduction of Perchlorate and Nitrate

Contents

17.1 EXPLOSIVES AND DECOMPOSING EXPLOSIVES

In this section, the explosive used in the September 17 bombing in 2016 at New York City and enzymes which can be used for cleaning up this explosive are explained. In the New York City attack, Tannerite (a commercially available explosive) and hexamethylene triperoxide diamine (HMTD; see Fig. 17.1.1A) were suspected to have been used. In the next section, decomposition of 2,4-dinitroanisole (DNAN; Fig. 17.1.1B) by a biological method is introduced.

17.2 NEW YORK CITY BOMBING, SEPTEMBER 2016

On September 17–19, 2016, three bombs exploded, and several explosives were found in New York City and near town, and 31 people were wounded. On the morning of September 17, a pipe bomb exploded in Seaside Park, New Jersey. A cooker bomb exploded in Chelsea later that

Biochemistry for Materials Science
https://doi.org/10.1016/B978-0-12-817054-0.00017-5
© 2019 Elsevier Inc.
All rights reserved.

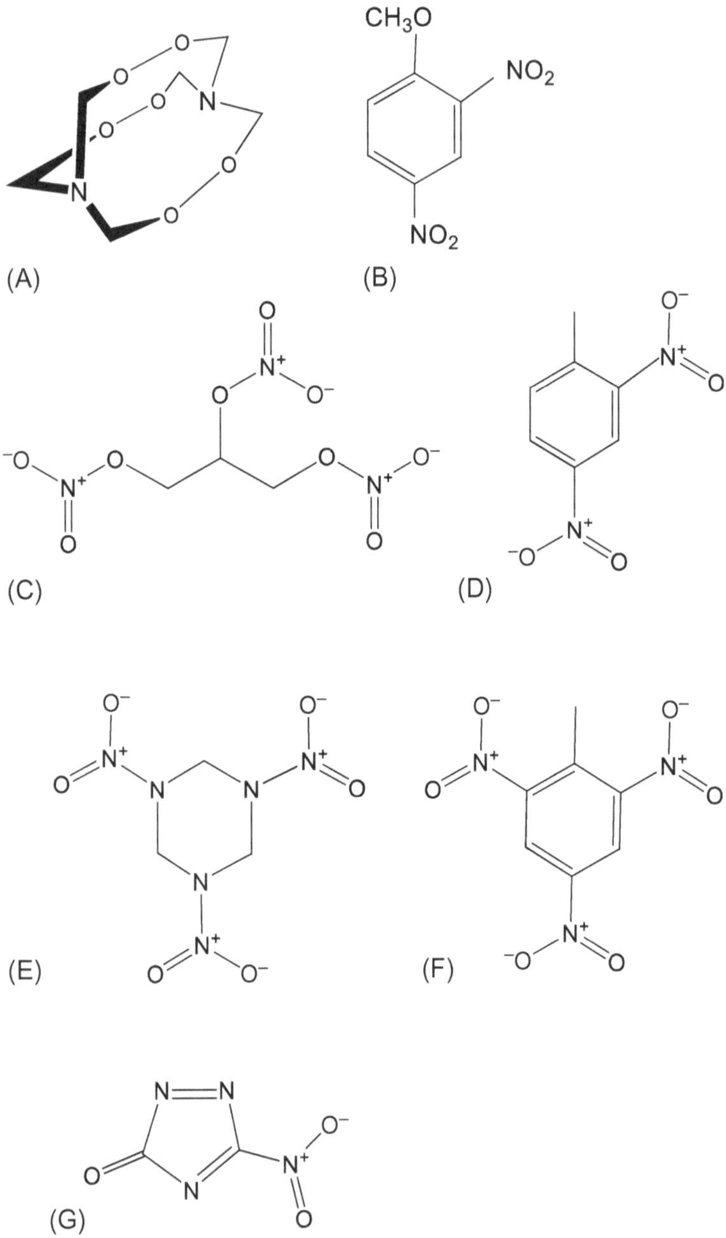

Fig. 17.1.1 Explosives. (A) Hexamethylene triperoxide diamine (HMTD), (B) 2,4-dinitroanisole (DNAN), (C) nitroglycerin, (D) dinitrotoluenes, (E) 1,3,5-trinitroperhydro-1,3,5-triazine (RDX), (F) 2,4,6-trinitrotoluene (TNT) and (G) 3-nitro-1,2,4-triazol-5-one (NTO).

day. The next day, several bombs were found in Elizabeth, New Jersey. On September 19, the sole suspect was captured.

Initially, a commercially available explosive called Tannerite was said to have been used. Later, it was confirmed that Tannerite alone could not have caused the explosion, and hexamethylene triperoxide diamine (HMTD; see Fig. 17.1.1A) was suspected to have been used as well.

Tannerite is made and sold by Tannerite Sports, which is used to explode targets in long-range shooting practice. When a target containing Tannerite is hit by a bullet, the target explodes, allowing shooters to hear and see more easily whether they shot successfully or not. From the Patent (Tanner, 2005), Tannerite is an 8:1 mixture by weight of ammonium nitrate and ammonium perchlorate, and a catalyst of 90% explosive grade aluminum powder, 5% titanium sponge, and 5% zirconium hydride. Tannerite is a very stable material, and cannot be exploded from being shot with the speed of 2000 km/s. Therefore, Tannerite needs a strong igniting shock, which can be given by HMTD. This was used in the 2015 Paris attacks and the 2016 Brussels bombings. However, HMTD is not a stable material, and explodes easily. Therefore, specialist knowledge of chemistry is required when handling it (Cross, 2016), whether for good or evil.

17.3 ENZYME-DECOMPOSING EXPLOSIVE

Military explosives were designed for performance, and little attention was paid to environmental effects or biological degradation in their development. Several common explosives, such as nitroglycerin (Fig. 17.1.1C; Husserl et al., 2010), dinitrotoluenes (Fig. 17.1.1D; Nishino et al., 2000), and 1,3,5-trinitroperhydro-1,3,5-triazine (RDX) (Fig. 17.1.1E; Binks et al., 1995) work as growth substrates for bacteria.

Recent development of insensitive munitions are 2,4-dinitroanisole (DNAN; Fig. 17.1.1B), 2,4,6-trinitrotoluene (TNT; Fig. 17.1.1F), and 3-nitro-1,2,4-triazol-5-one (NTO; Fig. 17.1.1G). DNAN is less sensitive to heat and shock than TNT, but is decomposable by bacteria (Karthikeyan et al., 2016); therefore DNAN is replacing TNT. However, DNAN contamination occurs in industrial wastewaters and soil, and firing sites.

A *Nocardioides* strain, JS1661, can decompose DNAN under various conditions (Karthikeyan et al., 2016). It can use carbon, nitrogen, and energy from solely DNAN. The key is a famous ether hydrolase or EC3.3.2.14, which is by Nomenclature Committee of the International Union of

Biochemistry and Molecular Biology (NC-IUBMB). The enzyme, which is a hexamer of two subunits, does not require external cofactors. This suggests that the enzyme works without the bacterial cells (Fida et al., 2014). Cell-free biocatalysts have the advantage of being able to be used in vitro (in test tubes). Biomimetic silica particles are synthesized in vitro in a process as diatoms, which incorporate biogenic silica as a scaffold in biosilicification.

The biosilicified enzyme was immobilized and packed in a microreactor consisting of a 2cm × 2mm stainless steel column (see Fig. 17.3.1) by Karthikeyan et al. (2016). The reactor was operated at room temperature. The reactor was connected to a biosensor system for 2,4-dinitrophenol (DNP), which was monitored at 400nm. The flow rate was maintained at 4μL/min, and the concentration of DNAN was 500μM. The enzyme was active over a pH range 6.2–9.5.

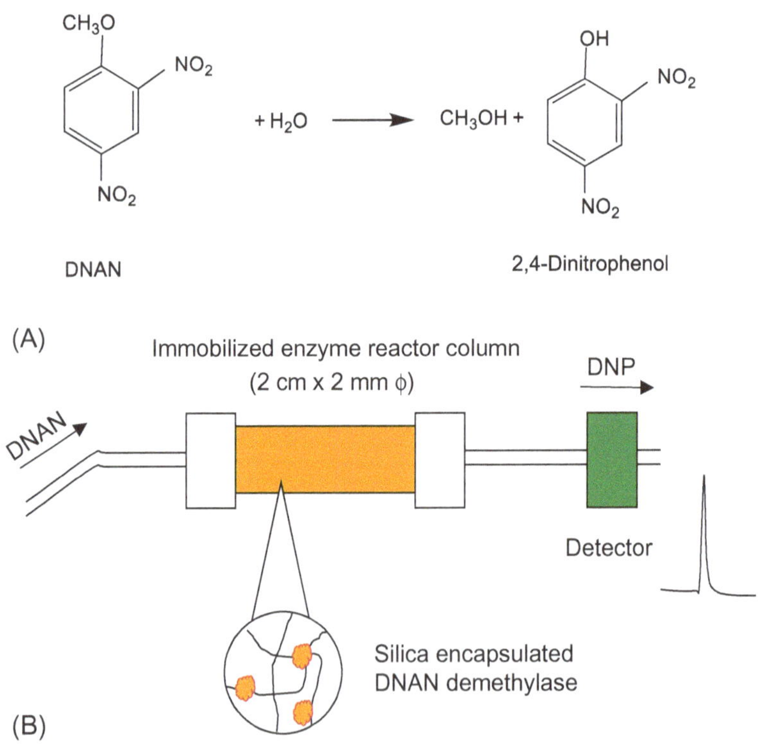

Fig. 17.3.1 (A) DNAN ether hydrolase (DNAN demethylase); EC 3.3.2.14 (IUBMB Nomenclature). (B) A schematic diagram of continuous decomposition system for DNAN.

Karthikeyan et al. (2016) also coated cellulose filter paper with the enzyme in order to monitor DNAN contamination in the field.

17.4 PERCHLORATE AND NITRATE REDUCTION

In Section 17.3, recent methods to decompose the explosives were shown. As pure acids, both perchloric acid and nitric acid are toxic and explosive. Their handling thus requires care. When heated, perchloric acid emits very explosive fumes. Nitric acid also emits lethal orange fumes when it reacts with reductive materials (Makishima, 2016). However, once they are diluted and neutralized, they become inorganic anions, such as perchlorate and nitrate anions, ClO_4^- and NO_3^-, respectively, which are difficult to decompose.

The catalytic reduction of these inorganic oxyanions is also difficult because of their weak complexation, poor nucleophilicity, and consequent inertness to oxidation. Therefore, severe reaction conditions such as low pH, high temperature, and long reaction times are required.

Despite their difficulty of decomposition, chlorine- and nitrogen-containing oxyanions are used in fertilizers, bleaching agents, and explosives. As they have high solubility and mobility in water, they have become contaminants in many sources of drinking water.

The most efficient reduction of chlorine- and nitrogen-containing oxyanions is achieved by the microbial metalloenzymes of (per)chlorate and nitrate reductases in vivo, respectively. Developments of their reduction methods in vitro using new catalysts are urgently required.

17.5 BIOLOGICAL OXYANION REDUCTION

Biological oxyanion reductions are explained in Fig. 17.5.1. The stabler oxyanions, ClO_4^-, ClO_3^-, and NO_3^- are reduced by perchlorate, chlorate, and nitrate reductases, respectively. The active sites of these three enzymes are Mo^{IV} complexes connected with S and O (see Fig. 17.5.1). The less stable oxyanions, ClO_2^- and NO_2^-, are reduced into $Cl^- + O_2$ and NO by chlorite dismutase and nitrite reductase, respectively. These two reductive reactions are performed by Fe^{II} stabilized in a heme complex (see Fig. 2.3.2), where the active sites of the metal are stabilized by four nitrogen with two carboxyl groups. In addition, another nitrogen of imidazole (Fig. 17.5.2) is connected to the metal.

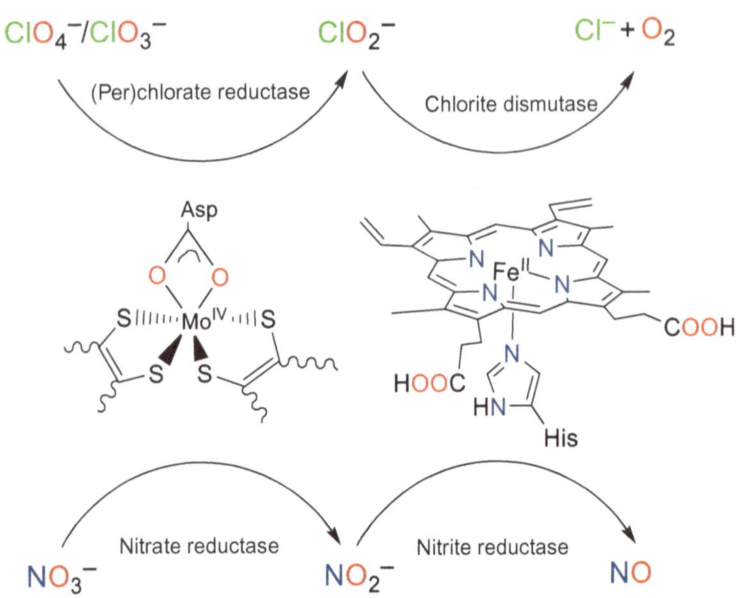

Fig. 17.5.1 Biological oxyanion reduction summarized by Ford et al. (2016).

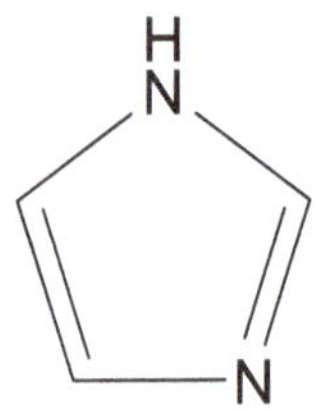

Fig. 17.5.2 Imidazole.

17.6 DEVELOPMENT OF A DEOXYGENATION CATALYST

Ford et al. (2016) synthesized a deoxygenation material by following the natural reductases: (i) a non-heme platform; (ii) an earth-enriched redox-active metal center (iron is best); (iii) a secondary coordination sphere that facilitates deoxygenation of substrates; and (iv) the ability of the ligand to undergo tautomerization, which facilitates proton and electron transfer between the substrate and the metal center.

Ford et al. reported the synthesis and characterization of transition-metal complexes of $[N(afa^{Cy})_3Fe^{II}OTf]OTf$ (Fe^{II}-OTf; afa= azafulvene-amine; Cy=cyclohexane; Tf= trifluoromethanesulfonate or $-SO_2-CF_3$; see

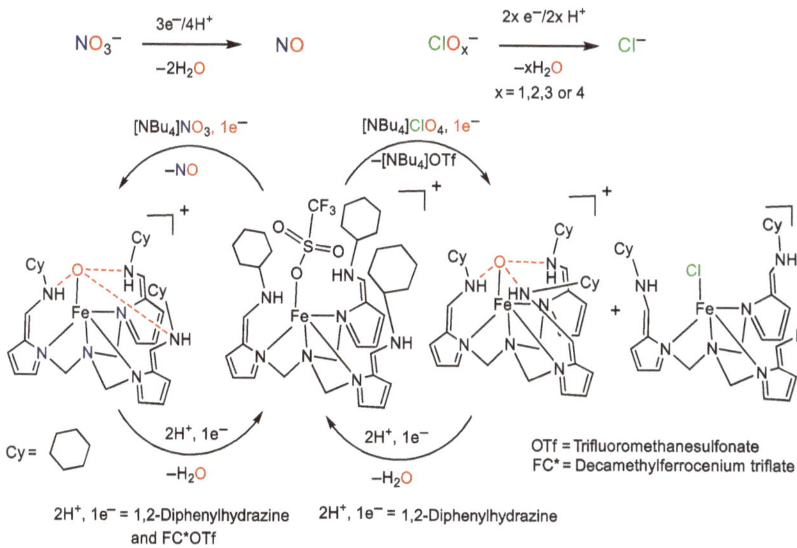

Fig. 17.6.1 Catalytic nitrate and perchlorate reduction developed by Ford et al. (2016).

Fig. 17.6.1). The ability of the ligand is key because it can facilitate proton and electron transfer between the substrate and the metal center. From Fig. 17.6.1, it can be seen that Fe is guarded by three afa^{Cy}, which makes it difficult for any ions to reach the Fe^{II} ion.

17.7 HOW DOES THE CATALYST WORK?

When the labile ClO_4^- was added as $[NBu_4]ClO_4^-$ with 5 eq of Fe^{II}-OTf with NEt_3, two products, Fe^{III}-O and $[N(afa^{Cy})_3Fe^{II}Cl]OTf$ (Fe^{II}-Cl) were detected in a 4:1 ratio. It was estimated that the perchlorate was stepwise reduced, from perchlorate to chlorate, chlorite, and hypochlorite.

When $[NBu_4]NO_2$ was added to 2 eq of Fe^{II}-OTf, $[N(afa^{Cy})_3Fe^{II}OTf]$ OTf, $[N(afa^{Cy})_3Fe^{III}O]OTf$ (Fe^{III}-O) and NO(g), which was trapped by Fe^{II}-OTf to give $[N(afa^{Cy})_3Fe^{II}NO]OTf_2$ (Fe^{II}-NO; the iron(II)-nitrosyl species), were formed. This is like nitrite reductase in nature.

When $[NBu_4]NO_3$ was added to 3 eq of Fe^{II}-OTf with presence of trimethylamine (NEt_3), 2/3 gave Fe^{III}-O, and 1/3 was Fe^{II}-NO. This can be easily reduced by one-electron of nitrite, which can be obtained from Fe^{II}-OTf. Thus, nitrate is reduced to nitrite in nature by these two-step reactions.

17.8 REGENERATION OF THE CATALYST

These authors investigated regeneration of Fe^{III}-O and Fe^{II}-Cl. The former was regenerated by 1,2-diphenylhydrazine (DPH) and Fc*OTf (decamethylferrocenium triflate in which all H of ferrocene (Fig. 1.2.3B) are methylated). The latter was regenerated with a mixture of Fe^{II}-OTf and DPH, which yielded azobenzene.

Thus, the Fe^{II}-OTf catalyst can reduce perchlorate and nitrate, and can be regenerated with simpler methods. Alternatively, the Fe^{II}-OTf catalyst can work with simultaneous regeneration.

REFERENCES

Binks, P.R., Nicklin, S., Bruce, N.C., 1995. Degradation of hexahydro-1,3,5-trinitro-1,3,5-triazine (RDX) by *Stenotrophomonas maltophilia* PB1. Appl. Environ. Microbiol. 61, 1318–1322.

Cross, R., 2016. Explosives used in New York bombing. Reports suggest bombs contained a combination of Tannerite and HMTD. C&EN 94, 7.

Fida, T.T., Palamuru, S., Pandey, G., Spain, J.C., 2014. Aerobic biodegradation of 2,4-Dinitroanisole by Nocardioides sp. strain JS1661. Appl. Environ. Microbiol. 80, 7725–7731.

Ford, C.L., Park, Y.J., Matson, E.M., Gordon, Z., Fout, A.R., 2016. A bioinspired iron catalyst for nitrate and perchlorate reduction. Science 354, 741–743.

Husserl, J., Spain, J.C., Hughes, J.B., 2010. Growth of *Arthrobacter* sp. strain JBH1 on nitroglycerin as the sole source of carbon and nitrogen. Appl. Environ. Microbiol. 76, 1689–1691.

Karthikeyan, S., Kurt, Z., Pandey, G., Spain, J.C., 2016. Immobilized biocatalyst for detection and destruction of the insensitive explosive, 2,4-dinitroanisole (DNAN). Environ. Sci. Technol. 50, 11193–11199.

Makishima, A., 2016. Thermal Ionization Mass Spectrometry (TIMS). Silicate Digestion, Separation, Measurement. Wiley-VCH Verlag GmbH & Co KGaA, Weinheim, ISBN: 978-3-527-34024-8p. 348.

Nishino, S.F., Paoli, G.C., Spain, J.C., 2000. Aerobic degradation of dinitrotoluenes and pathway for bacterial degradation of 2,6-dinitrotoluene. Appl. Environ. Microbiol. 66, 2139–2147.

Tanner, D.J. 2005. Binary exploding target, package process and product. US patent, US6848366 (B1).

CHAPTER EIGHTEEN

Teeth- and Bone-Related Topics and Materials

Contents

18.1 ENHANCING THE LIFE OF DENTAL FILLINGS BY PINE BARK EXTRACT

Dentin is made of the protein collagen. It composes the bulk of the calcified extracellular tissue of teeth. Dentin forms most of the layer beneath the hard outermost enamel.

Bedran-Russo et al. (2014) found that extracts from pine bark (*Pinus massoniana*) form chemical cross-links with collagen in dentin, and that the extracts enhance the hardness of dentin by 10 times; the dentin-pine bark extracts cross-links are called as the dentin-enhancing oligomeric proanthocyanidins (OPACs). Bedran-Russo et al. found proanthocyanidins, which are flavonoids in many natural materials and exist as oligomers of up to six or more linked subunits. The proanthocyanidins were determined to work as the active components of the bark extracts by Bedran-Russo et al. (2014).

Resin-based composites are used as tooth-colored fillings, which repair the surface of teeth beautifully. However, they often break after 5–8 years. Nam et al. (2017), who are in the same group of Bedran-Russo et al. (2014),

https://doi.org/10.1016/B978-0-12-817054-0.00018-7
© 2019 Elsevier Inc.
All rights reserved.

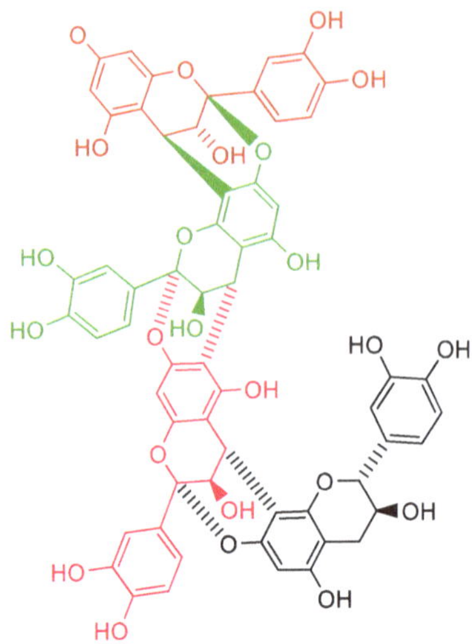

Fig. 18.1.1 A tetrameric proanthocyanidin.

characterized the OPACs as strengthening the interface between a tooth and the resin-fillings, and as extending the life span of the dental fillings.

Nam et al. (2017) separated OPACs into fractions using chromatography, and tested each fraction for dentin-strengthen activity. They applied the fractions to pieces of a human molar, and measured deformation of the tooth after applying a force. They found the fractions containing trimeric and tetrameric proanthocyanidins showed the best performance. They then characterized the structures using NMR spectroscopy (see Fig. 18.1.1 for a tetrameric proanthocyanidin).

18.2 BIOCOMPATIBLE BINDER FOR BONE GRAFTS

In order to repair bone tissue, doctors sometimes use bone materials taken from other parts of a patient's body. However, taking healthy bone tissue can be very painful, and the pain can last for a long time in the spot from where the tissue was taken. Bone materials from cattle are safe and abundant and can avoid these complications for humans; however, the materials can trigger an immune response in humans. Therefore, collagen and other structural proteins in the bone must be removed. At present, such processes make the material brittle. Materials such as cellulose gum can restore some of the lost cohesion, but these binders can lose their function in the blood and water of the human body.

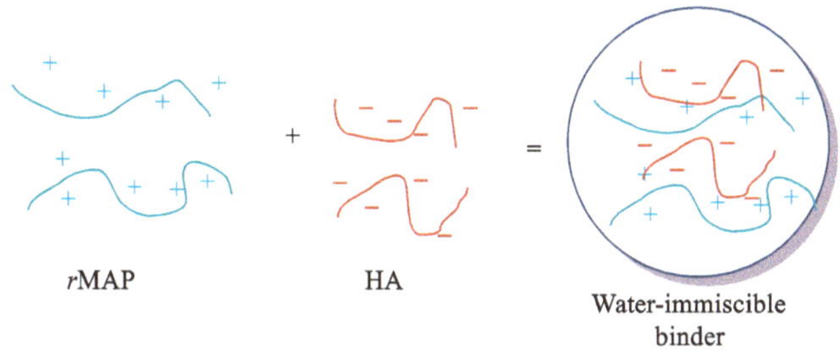

Fig. 18.2.1 A biocompatible binder complex made of positively charged recombinant mussel adhesive protein (rMAP) and negatively charged hyaluronic acid (HA).

In order to solve these problems, Kim et al. (2016), inspired by a sandcastle worm coacervating complex, found a biocompatible binder based on two polyelectrolytes: negatively charged hyaluronic acid (HA) and positively charged recombinant mussel adhesive protein (*r*MAP). Bovine bone particles treated with the *r*MAP/HA coacervate binder (see Fig. 18.2.1) remained stable and sticky even in blood and water. Furthermore, studies in rats showed that the implanted material promoted bone regeneration.

18.3 HOW DOES THE NACRE (MOTHER OF PEARL) GROW?

Aragonite and calcite are the two calcium carbonate polymorphs that give the shell of molluscan bivalves their strength and elasticity, due to the nano- and microstructural assembly of the overall architecture.

Nacre, or mother of pearl, has not yet been identified as osteoinductive or osteointegrative. It is composed of aragonite and induces production of vertebrate bone (calcium phosphate). This remarkable phenomenon has been confirmed in vitro and in vivo, yet the characteristic of nacre that induces bone formation remains unknown (Westbrock and Marin, 1998).

By isolating nacre topography from its inherent chemistry in the production of polycaprolactone (PCL) nacre replica, Alakpa et al. (2017) showed that for mesenchymal stem cells (MSCs), nacre topography is osteoinductive. Gene expression of specific bone marker proteins osteopontin, osteocalcin, osteonectin, and osterix was increased 10-, 2-, 1.7-, and 1.8-fold, respectively, when compared to planar PCL. Furthermore, Alakpa et al. (2017) demonstrated that bone tissue formed in response to the topographical features of nacre has a higher crystallinity than bone formed in response to chemical cues. These differences in mineral product are supported by differences in cellular metabolism. This observation can be used in the design of bone therapies, which is most pressing in light of a rapidly aging human population.

18.4 COLLAGEN IN DINOSAUR BONE

18.4.1 Introduction

Organic components of animals decompose during the fossilization process. Minerals in groundwater are redeposited in the interstitial spaces. Finally, vertebrate bones and teeth are transformed into inorganic "rock" (Schweitzer, 2011). However, some fossils are preserved so exceptionally well that even soft tissue is maintained when the fossils are isolated and protected by an inorganic environment. Such cases attract a lot of attention because of the valuable biological information that may be obtained. In particular, the preservation of biomolecules in cells and blood vessels was found in Cretaceous dinosaurs. The analyses of their chemical composition were attempted using scanning electron microscopy (SEM), transmission electron microscopy (TEM), immunological assays, and mass spectrometry (MS) (Schweitzer et al., 2007a, b). These studies did not provide clear and direct in situ evidence for entire molecules of protein or collagen by exogenous contaminations, such as bacteria biofilm and hematite (α-Fe_2O_3), which are clarified by synchrotronradiation-based Fourier transform infrared (SR-FTIR) microspectroscopy and scanning energy dispersive spectroscopy of scanning electron microscopy (SEDS-SEM), respectively. Nevertheless, recent studies have supported the idea that the degradation structure of collagen, Gly-Pro-Hyp (Glycine-Proline-hydroxyProline) sequence, was observed using MS and preserved from 75 Mya to the present (Bertazzo et al., 2015). The evidence of Gly-Pro-Hyp sequence is matched with the general collagen sequence Gly-X-Y, where the X and Y positions are proline and hydroxyproline, respectively.

18.4.2 Evidence of Protein Preservation in 195-Million-Year-Old Dinosaurs

Lee et al. (2017) identified the organic materials preserved in much older fossils using in situ microspectroscopic methods in mid-Mesozoic terrestrial vertebrates. Various skeletal elements of the sauropodomorph dinosaur *Lufengosaurus* from the Lower Jurassic (Early Jurassic, Sinemurian, 190–197 Mya) of Dawa, Lufeng County, Yunnan Province, China, were studied in detail (e.g., Reisz et al., 2013); the rib materials showed the most significant results. Adult compact bone (including ribs) is composed of osteons (the fundamental functional unit of compact bone) with central vascular canals that contained blood vessels and nerves in the living organism and lacunae (spaces) for the adult osteocytes.

In order to exclude the possibility of external contamination, in situ investigation of organic remains within the adult *Lufengosaurus* bones was undertaken using thin sections. FTIR spectroscopy is a nondestructive analysis method for identifying the vibrational motions of chemical bonding of molecular structures, based on the characteristic infrared absorption bands in the mid-infrared range, corresponding to the various functional groups, especially for organic molecules including proteins (Lebon et al., 2011; Jadwiszczak et al., 2013; Brink et al., 2015; Lang et al., 2016; Surmik et al., 2016). Typically, the organics remaining within fossils are rare, and are distributed within a tiny space at the micrometer scale. In situ SR-FTIR microspectroscopy was used to obtain not only ultrahigh laterally resolved FTIR spectra but also chemical images related to chemical bonding of molecules in a diffraction-limited area to reveal the organics remaining within the fossil bone. Raman microscopy is a complementary tool to SR-FTIR microspectroscopy and accessible to give low vibrational frequencies of chemical bonding of molecules, particularly for metal oxide compounds in the far-infrared range (Thomas et al., 2007; Schopf and Kudryavtsev, 2009), which are not obtained by FTIR microspectroscopy.

Lee et al. (2017) used dark-field microscopy, synchrotron radiation-based nano-transmission X-ray microscopy (SR-TXM; Yin et al., 2006; Wang et al., 2015), and transient absorption microscopy (TAM; Jung et al., 2010; Villafana et al., 2015) to reveal three-dimensional (3D) lacunae within the fossil bone, and two-dimensional (2D) transient absorption images of hematite distributed in the vascular canals and lacunae.

The results indicate that native collagen was preserved within the osteonal central vascular canals, together with hematite particles (6–8 μm in diameter) that were possibly derived from hemoglobin of the dinosaur. This suggests that hemoglobin and the resulting hematite may have had a preservative effect on the proteins that originally resided in the central osteonal canals of the rib.

18.5 SYNTHETIC TOOTH ENAMEL

Tooth enamel is made of parallel microscale and nanoscale ceramic columns or prisms filled with a soft protein matrix. This structure is similar in all species across all geological times. This invariability suggests the existence of a functional basis.

Yeom et al. (2017), inspired by the enamel structure, made ex vivo replicate columnar nanocomposites by sequential growth of zinc oxide (ZnO) nanowire carpets followed by layer-by-layer (LBL) deposition of a polymeric matrix around the carpets. The LBL matrix was made of polyallylamine (PAAm) and polyacrylic acid (PAA). Finally the nanocomposites had a $(ZnO/LBL)_n$ structure.

Yeom et al. (2017) showed that the mechanical properties of these nanocomposites (abiotic enamels; AEs) are comparable to those of natural tooth enamels. The resistance to vibrational damage of AEs is higher than the tooth enamels. The high stiffness, damping, and light weight in these layer-by-layer AEs are based on efficient energy dissipation in the interfacial portion of the organic phase.

The in vivo contribution of this interfacial portion to macroscale deformations along the natural tooth enamels is maximized when the architecture is columnar. This suggests an evolutionary advantage of the columnar motif in the enamel of living species.

Their findings should be applied to all columnar composites and to the development of high-performance load-bearing materials.

REFERENCES

Alakpa, E.V., Burgess, K.E.V., Chung, P., Riehle, M.O., Gadegaard, N., et al., 2017. Nacre topography produces higher crystallinity in bone than chemically induced osteogenesis. ACS Nano 11, 6717–6727.

Bedran-Russo, A.K., Pauli, G.F., Chen, S.-N., McAlpine, J., Castellana, C.S., et al., 2014. Dentin biomodification: strategies, renewable resources and clinical applications. Dent. Mater. 30, 62–76.

Bertazzo, S., et al., 2015. Fibres and cellular structures preserved in 75-million-year-old dinosaur specimens. Nat. Commun. 6, 7352.

Brink, K.S., et al., 2015. Developmental and evolutionary novelty in the serrated teeth of theropod dinosaurs. Sci. Rep. 5, 12338.

Jadwiszczak, P., et al., 2013. The first record of fossil penguins from East Antarctica. Antarct. Sci. 25, 397–408.

Jung, Y., et al., 2010. Fast detection of the metallic state of individual single-walled carbon nanotubes using a transient-absorption optical microscope. Phys. Rev. Lett. 105, 217401.

Kim, H.J., Choi, B.-H., Jun, S.H., Cha, H.J., 2016. Sandcastle worm-inspired blood-resistant bone graft binder using a sticky mussel protein for augmented in vivo bone regeneration. Adv. Healthc. Mater. https://doi.org/10.1002/adhm.201601169.

Lang, L., Kirsimä, K., Vahur, S., 2016. Diagenetic fate of bioapatite in linguliform brachiopods: multiple apatite phases in shells of Cambrian lingulate brachiopod ungula ingrica (Eichwald). Lethaia 49, 13–27.

Lebon, M., et al., 2011. Imaging fossil bone alterations at the microscale by SR-FTIR microspectroscopy. J. Anal. At. Spectrom. 26, 922–929.
Lee, Y.-C., Chiang, C.-C., Huang, P.-Y., Chung, C.-Y., Huang, T.D., et al., 2017. Evidence of preserved collagen in an early Jurassic sauropodomorph dinosaur revealed by synchrotron FTIR microspectroscopy. Nat. Commun. 8, 14220. https://doi.org/10.1038/ncomms14220.
Nam, J.-W., Phansalkar, R.S., Lankin, D.C., McAlpine, J.B., Leme-Kraus, A.A., et al., 2017. Absolute configuration of native oligomeric proanthocyanidins with dentin biomodification potency. J. Org. Chem. 82, 1316–1329.
Reisz, R.R., et al., 2013. Embryology of early Jurassic dinosaur from China with evidence of preserved organic remains. Nature 496, 210–214.
Schopf, J.W., Kudryavtsev, A.B., 2009. Confocal laser scanning microscopy and Raman imagery of ancient microscopic fossils. Precambrian Res. 173, 39–49.
Schweitzer, M.H., et al., 2007a. Analyses of soft tissue from *Tyrannosaurus rex* suggest the presence of orotein. Science 316, 277–280.
Schweitzer, M.H., Wittmeyer, J.L., Horner, J.R., 2007b. Soft tissue and cellular preservation in vertebrate skeletal elements from the cretaceous to the present. Proc. R. Soc. B 274, 183–197.
Schweitzer, M.H., 2011. Soft tissue preservation in terrestrial Mesozoic vertebrates. Annu. Rev. Earth Planet. Sci. 39, 187–216.
Surmik, D., et al., 2016. Spectroscopic studies on organic matter from Triassic reptile bones, upper Silesia, Poland. PLoS ONE. 11,e0151143 (2016).
Thomas, D.B., Fordyce, R.E., Frew, R.D., Gordon, K.C., 2007. A rapid, non-destructive method of detecting diagenetic alteration in fossil bone using Raman spectroscopy. J. Raman Spectrosc. 38, 1533–1537. (2007).
Villafana, T.E., Brown, W., Warren, W.S., Fischer, M., 2015. Ultrafast pump-probe dynamics of iron oxide based earth pigments for applications to ancient pottery manufacture. In: Proc. SPIE 9527, optics for arts, architecture, and archaeology V, 952709 (SPIE, Munich, Germany, 2015).
Wang, C.-C., et al., 2015. Evolution and function of dinosaur teeth at ultramicrostructural level revealed using synchrotron transmission X-ray microscopy. Sci. Rep. 5, 15202.
Westbroek, P., Marin, F., 1998. A marriage of bone and nacre. Nature 392, 861–862.
Yeom, B., Sain, T., Lacevic, N., Bukharina, D., Cha, S.-H., et al., 2017. Abiotic tooth enamel. Nature 543, 95–99.
Yin, G.-C., et al., 2006. Sub-30 nm resolution X-ray imaging at 8 keV using third order diffraction of a zone plate lens objective in a transmission microscope. Appl. Phys. Lett. 88, 241115.

CHAPTER NINETEEN

Flu Virus and Chemical Detectors, and Diagnostics

Contents

19.1 DETECTION OF THE FLU VIRUS AT THE DOCTOR'S OFFICE

According to the World Health Organization (WHO), influenza infects 0.4–0.6 billion children and 0.2–5.0 billion adults worldwide in 2016. Emerging pandemic strains can infect a large number of people rapidly. For example, the H1N1 2009 "swine" strain pandemic spread rapidly to over 74 countries in less than two months. While influenza vaccines are particularly recommended for the most susceptible groups, only ~70% of people over the age of 65 and 50% of children aged 6–18 received vaccines in the United States for the 2015–16 season. Therefore, other countermeasures including rapid diagnostics are urgently required.

The typical detection methods for influenza viruses are viral isolation and growth using standard cell culture, nucleic acid identification, and

Biochemistry for Materials Science
https://doi.org/10.1016/B978-0-12-817054-0.00019-9
© 2019 Elsevier Inc.
All rights reserved.

antibody-based immunoassays. Cell culture and nucleic acid identification are expensive, require trained personnel, and are not suited for financially poor patients. Antibody-based immunochromatography tests are user-friendly, but have low clinical sensitivity, especially at low pathogen counts. These strip tests are typically used to "rule out" the infection, and require further tests.

An ideal biosensor should meet the criteria of ASSURED (Affordable, Sensitive, Selective, User-friendly, Rapid and robust, Equipment-free, and Deliverable to end-users) that are particularly important for fast-spreading respiratory infections. If diagnostics are cheap and user-friendly, more people will participate in monitoring, resulting in faster countermeasures being achieved.

Cui et al. (2017) recently developed an electrochemical assay to detect influenza viruses using a glucose-bearing substrate (Zhang et al., 2015). The neuraminidase (NA) enzyme is present in high abundance on the surface of influenza viruses (Harris et al., 2006). The NA enzyme cleaves the substrate to release glucose, which is detected perimetrically by a glucose meter (see Figs. 19.1.1 and 19.1.2). Glucose meters also agree with the ASSURED criteria and are used by millions worldwide.

Glucose meters are used to detect a variety of nonglucose targets (Mohapatra and Phillips, 2013; Chavali et al., 2014; Wang et al., 2015; Gurale et al., 2016). In a previous report by the same group, SG1

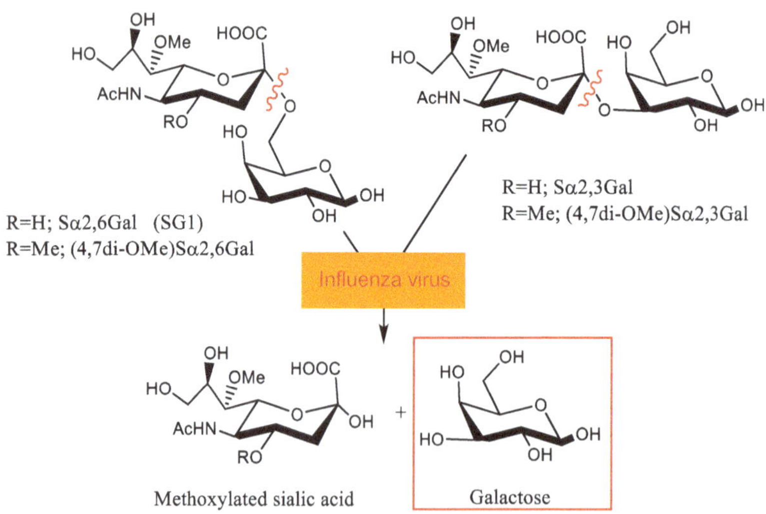

Fig. 19.1.1 Neuramidaze enzyme of influenza virus selectively decomposes glycans above into methoxylated sialic acid and galactose.

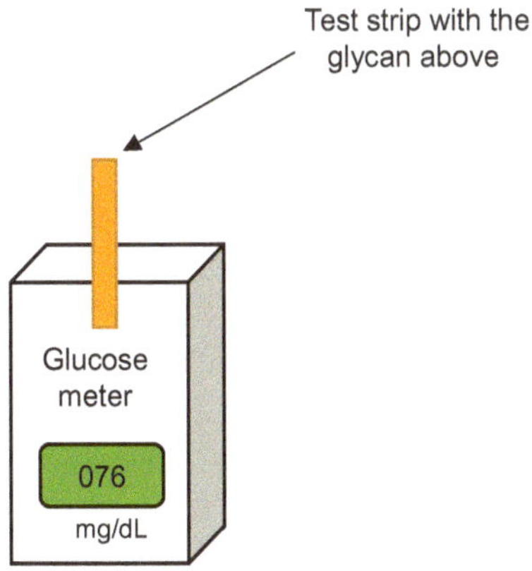

Fig. 19.1.2 A hand-held glucose meter detects the liberated galactose.

(see Fig. 19.1.1; Sα2,6Gal or Sα2,3Gal in which R = H) was used to detect 19 unique influenza strains in 1 h. However, SG1 is cleaved by NA from any source (human, bacterial, or viral), which makes separation of influenza virus from other pathogens very difficult without an additional step of using antiviral drugs.

Thus, Cui et al. (2017) redesigned the compounds. Influenza surface glycoprotein neuraminidase (NA) cleaves galactose-bearing substrates, 4,7di-OMe *N*-acetylneuraminic acid attached to the 3 or 6 positions of galactose (see Fig. 19.1.1; (4,7di-OMe)Sα2,6Gal or (4,7di-OMe)Sα2,3Gal in which R = CH_3), to release galactose. In contrast, bacterial NA cannot cleave the natural substrate, 4,7di-OMe *N*-acetylneuraminic acid, attached to the 3 or 6 positions of galactose. Therefore, the detection of influenza virus could be performed with higher accuracy. Cui et al. used portable potentiated and commercial Accu-Chek test strips that employ dehydrogenase enzymes to detect galactose or glucose (see Fig. 19.1.2).

Furthermore, Cui et al. demonstrated the detection of clinically significant amounts of influenza virus in spiked human nasal swabs, which indicated that this assay can be used to detect influenza in a clinical or home setting. In addition, the selectivity for influenza was confirmed using influenza strains, different respiratory pathogens such as *Streptococcus pneumoniae* and *Haemophilus influenza* (which is not a virus), and bacteria. Viral drug susceptibility and testing with human clinical samples was successful in 15 min, indicating that this assay could be used to detect influenza viruses quickly at primary care or resource-poor settings that had glucose meters, such as in a doctor's office.

19.2 ENGINEERED BACTERIA AS LIVE DIAGNOSTICS IN THE MAMMALIAN GUT

Bacteria can be engineered to function as diagnostics in the mammalian gut. Riglar et al. (2017) report stable, engineered bacterial strains that maintain their function for 6 months in the mouse gut. They engineered a commensal murine *Escherichia coli* strain to detect tetrathionate, which is produced during inflammation.

Using the engineered diagnostic strain, which retains memory of exposure in the gut for analysis by fecal testing, Riglar et al. detected tetrathionate in both infection-induced and genetic mouse models of inflammation over 6 months. The synthetic genetic circuits in the engineered strain were genetically stable and worked as intended over time.

The performance of the strains confirms the potential of engineered bacteria as live diagnostics.

19.3 APTAMER-BASED SENSORS IN LIVE ANIMALS

If the levels of drugs, metabolites, and biomarkers in the body could be traced continuously and in real time, this would advance our ability to detect and treat disease. For example, therapies by high-resolution, patient-specific dosing would open up new opportunities for personalized medicine.

Arroyo-Currás et al. (2017) developed electrochemical aptamer-based (*E*-AB) sensors (see Fig. 19.3.1) that enable continuous, real-time,

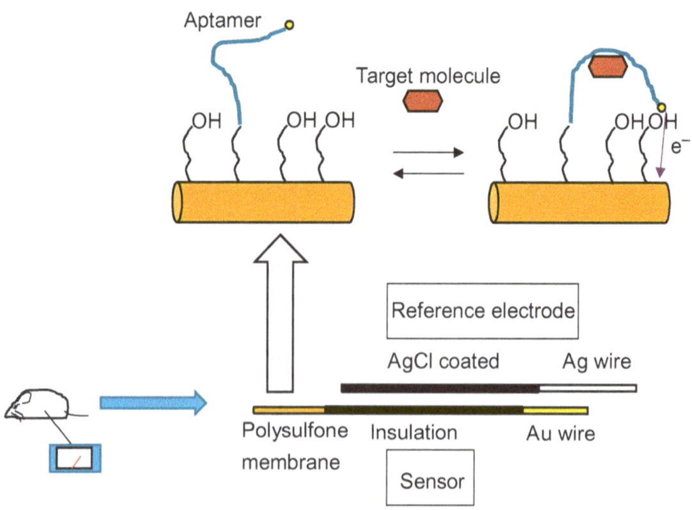

Fig. 19.3.1 Aptamer-based sensors in live animals.

multihour measurements when placed directly in the circulating systems of living animals. Aptamers are oligonucleotide or peptide molecules that bind to a specific target molecule. Arroyo-Currás et al. (2017) used the E-AB sensors to measure four drugs – doxorubicin (an anticancer drug), kanamycin (an antibiotic), gentamicin (an antibiotic), and tobramycin (an antibiotic)—in the bloodstream of conscious, ambulatory rats over multiple hours and in real time. They achieved precise molecular measurements of clinically important detection limits and high (3 s) temporal resolution.

This approach should provide an important window into the study of physiology and pharmacokinetics.

19.4 ON-SITE FOOD ALLERGEN DETECTION

Adverse food reactions such as food allergies affect 5%–15% of the population, and are a considerable public health problem.

Avoiding problematic foods is difficult, practically speaking, when prepared foods or out-of-home meals are eaten. To overcome this problem, Lin et al. (2017) developed a portable detection technology: integrated exogenous antigen testing (iEAT).

The system (Fig. 19.4.1) consists of a pocket-sized detector (Fig. 19.4.2), an electrode chip, and a disposable kit for allergen extraction. Antigen extraction is shown in Fig. 19.4.1B. Antigens are captured on magnetic beads (MBs) and then labeled with antibodies with oxidizing agents (horseradish peroxidase or HRP). A disposable kit has been developed. The magnet collects and disperses MBs. Then, HRP-coated MBs are mixed with electron mediators (TMB, 3,3′,5,5′- tetramethylbenzidine) and dropped on the electrode (Fig. 19.4.1C). HPR catalyzes the oxidation of TMB. Electrical currents are generated when oxidized TMB is reduced on or near the electrode.

Fig. 19.4.2 shows the schematic of an iEAT detector (reader). This pocket-sized reader is designed for stand-alone operation with display,

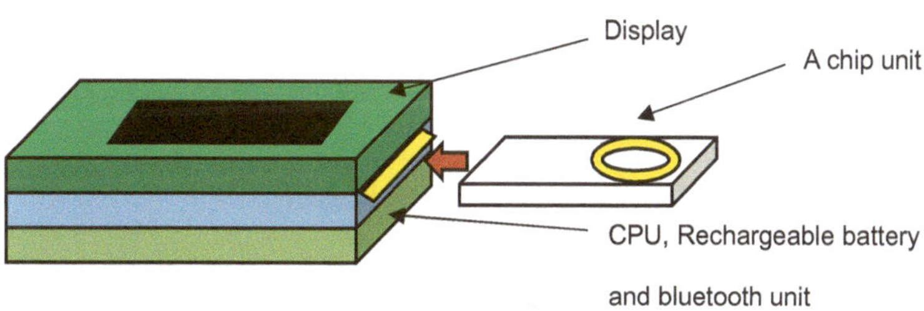

Fig. 19.4.2 Schematic of iEAT detector (reader).

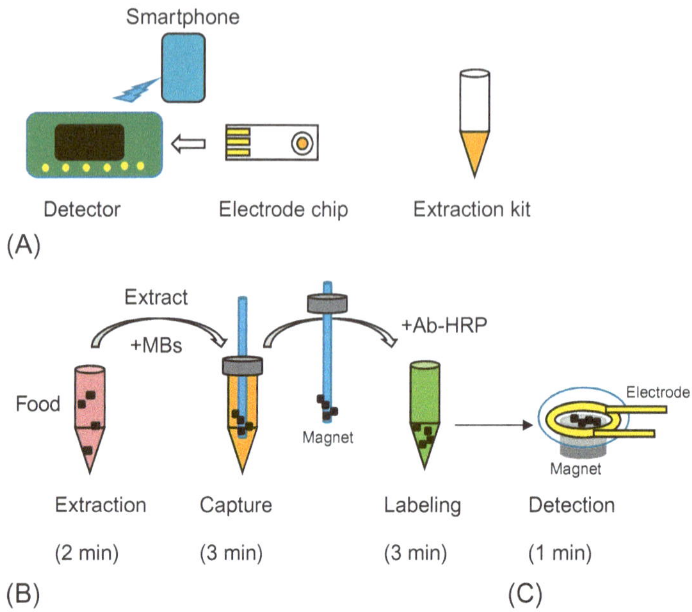

Fig. 19.4.1 iEAT system for onsite antigen detection developed by Lin et al. (2017). (A) The system consists of a pocket-sized detector, an electrode chip, and a disposable kit for allergen extraction. The detector connects with a smartphone for system control and data upload. (B) Antigen extraction. (C) Signal detection.

CPU, rechargeable battery, and wireless communication (e.g., Bluetooth unit) modules. A chip unit with a single electrode or multiple electrodes can be connected to the reader through a connector.

The prototype iEAT system was designed to detect five major food antigens in peanuts, hazelnuts, wheat, milk, and eggs. Antigen extraction and detection with the iEAT system requires <10 min and achieves high-detection sensitivities (e.g., $0.1\,mg\,kg^{-1}$ for gluten, which is lower than the regulatory limits of $20\,mg\,kg^{-1}$).

The small, rapid, and simple iEAT system should help not only consumers but also other key stakeholders such as clinicians, food industries, and regulators to enhance food safety.

19.5 BREATHALYZER-LIKE NANOARRAY DISEASE SENSOR

Since the time of the Ancient Greeks, physicians have learned to evaluate patients by exhaled volatile organic compounds (VOCs). Detection of diseases from exhaled breath is a low-cost, low-energy, and easily repeated method.

Recognition of VOCs by nanotechnology-based sensors can be achieved by selective detection of identified VOCs. This approach is useful to detect specific well-defined VOCs in the presence of interfering gaseous species. Even though advances have been made in detection of VOCs by highly selective nanomaterial-based recognition methods, most diseases today cannot be identified properly by individual VOCs alone.

Nakhleh et al. (2017) have developed an artificially intelligent nanoarray based on molecularly modified gold nanoparticles and a random network of single-walled carbon nanotubes for noninvasive diagnosis and classification of a number of diseases from exhaled breath.

The performance of this artificially intelligent nanoarray was clinically assessed on breath samples collected from 1404 subjects having one of 17 different disease conditions: chronic kidney failure (CKD), idiopathic Parkinson's disease (IPD), atypical Parkinsonism (PDISM), multiple sclerosis (MS), Crohn's disease (CD), ulcerative colitis (UC), irritable bowel syndrome (IBS), pulmonary arterial hypertension (PAH), pre-eclampsia in pregnant women (PET), head and neck cancer (HNC), lung cancer (LC), colorectal cancer (CRC), bladder cancer (BC), kidney cancer (KC), prostate cancer (PC), gastric cancer (GC), and ovarian cancer (OC), or having no evidence of any disease (healthy controls).

Blind tests showed that accuracy of 86% was achieved with the artificially intelligent nanoarray, allowing both detection and discrimination between the different disease conditions examined.

Analysis of the artificially intelligent nanoarray also showed that each disease has its own unique "breathprint," and that the presence of one disease would not screen out others. Cluster analysis showed a reasonable classification power of diseases from the same categories (see Fig. 19.5.1). The effect of confounding clinical and environmental factors on the performance of the nanoarray did not significantly alter the obtained results.

The diagnosis and classification power of the nanoarray were also validated by an independent analytical technique: gas chromatography linked with mass spectrometry. This analysis found that 13 exhaled chemical species (VOCs), are associated with certain diseases, and the composition of this assembly of VOCs differs from one disease to another.

In summary, these findings could contribute to one of the most important criteria for successful health intervention in the modern era, namely, easy-to-use, inexpensive, and miniaturized tools that could also be used for personalized screening, diagnosis, and follow-up of a number of diseases.

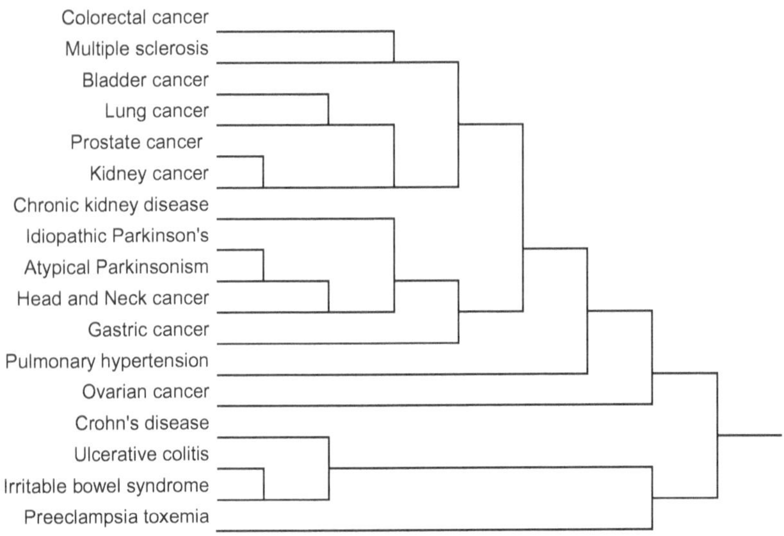

Fig. 19.5.1 Results of clustering analysis of the responses of the sensors. Each cluster represents a similar response profile, suggesting resemblance between subjects in a specific cluster.

19.6 SWEETENER TRACKS URINARY INPUT INTO SWIMMING POOLS

19.6.1 Introduction

The overnight color change of the water from blue to green in the 2016 Rio Olympic pools (Hutchins, 2016) highlighted the necessity of monitoring water quality in swimming pools. Various chemicals are introduced into recreational waters via bodily fluids, which can react with disinfectants. Recently, more than 100 disinfection byproducts (DBPs) in swimming pools and hot tubs have been found. Organic extracts in these samples were more mutagenic than corresponding tap water extracts (Daiber et al., 2016). Epidemiological studies have found a potential risk of bladder cancer associated with long-term DBP exposure via drinking water (Villanueva et al., 2004); however, association via exposure through swimming pools has been inconsistent.

Human urinary input into swimming pools is a public health concern, although urine itself is not toxic. Urine contains many nitrogenous compounds such as urea, ammonia, amino acids, and creatinine. These compounds can react with disinfectants such as chlorine in swimming pools to form DBPs, including trihalomethanes, haloacetic acids, haloamines,

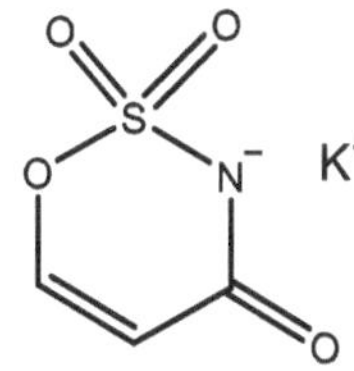

Fig. 19.6.1 Acesulfame-potassium (ACE).

and halonitromethanes. Exposure to volatile DBPs, specifically trichloramine, in indoor swimming facilities can lead to eye and respiratory irritation (Schmalz et al., 2011) and asthma (Thickett et al., 2002).

The average urine output per swimmer in pools is approximately 70 mL. Dissolved organic carbon (DOC) in swimming pools is related to formation of trihalomethanes. The potential negative health effects associated with DBPs led researchers to investigate a marker for urine in swimming pools.

Artificial sweeteners are consumed in large quantities because of their negligible calories and low impact on blood sugar (Gougeon et al., 2004). Nonnutritive artificial sweeteners such as acesulfame-K (ACE; see Fig. 19.6.1; Buerge et al., 2009), sucralose (SUC), saccharin (SAC), and cyclamate (CYC) have been used as indicators of wastewater (Van Stempvoort et al., 2011).

ACE is used in prepackaged foods, and is not metabolized by humans. ACE is completely excreted in the urine (Renwick, 1986), whereas SUC is excreted in feces. The average concentration of ACE in urine is approximately 4000 ng/mL (Zhang et al., 2016). ACE is stable at varying pHs and high temperatures, and is resistant to microbial action in aerobic soils. Furthermore, ACE is much more resistant to wastewater treatment processes than SAC or CYC. As ACE is widely used, stable, and persistent, it can be used as an indicator of urinary input in swimming pools.

Blackstock et al. (2017) started a new approach to determine urinary input from ACE. The objective of their study was to determine the occurrence of the artificial sweetener ACE in swimming pools compared to input tap water. ACE in water was directly determined by a high-throughput method using high-performance liquid chromatography (HPLC) with tandem mass spectrometry (MS/MS).

19.6.2 Urinary Input Into Swimming Pools

Blackstock et al. (2017) collected 250 samples from water in 31 pools, with corresponding tap water samples.

Concentrations of ACE ranged from 30 to 7110 ng/L, up to 570-fold greater than in the input tap water. The level of dissolved organic carbon was greater in all pools than in the input water. As a Canadian human urine sample ($N=20$) contains the average ACE of 2360 ng/mL, and the levels of ACE over 3 weeks in two pools (420,000 and 840,000 L) were 156 and 210 ng/L, the average urine contributions were calculated to be 30 and 75 L, respectively. Thus, this study clearly shows the human impact in pools and tubs.

In addition, this study is useful for future studies of the human contribution to DBP formation, epidemiological assessment of exposure, and unfavorable health effects in recreational facilities.

19.7 PHARMACOKINETICS OF BPA FROM THERMAL PAPER RECEIPTS

Bisphenol A (BPA) (see Fig. 19.7.1A) is an endocrine disruptor often detected in human fluids. Dermal absorption of BPA from thermal paper receipts occurs. However, dermal pharmacokinetics of BPA is not well understood.

Liu and Martin (2017) clarified the pharmacokinetics of dermal BPA exposure. They chose six male volunteers. The participants handled simulated receipts containing relevant levels of BPA (isotope-labeled BPA-d_{16}; see Fig. 19.7.1B) for 5 min, followed by hand-washing 2 h later. Free and total levels of BPA-d_{16} in urine (0–48 h) and serum (0–7.5 h) were monitored. One week later, participants returned for a dietary administration with monitoring as above.

After dermal exposure, cumulative excretion increased linearly for 2 days, and half the participants still had detectable urinary total BPA-d_{16} after 1 week. For comparison, dietary exposure for the participants were determined by Liu and Martin (2017). Urine total BPA-d_{16} peaked within

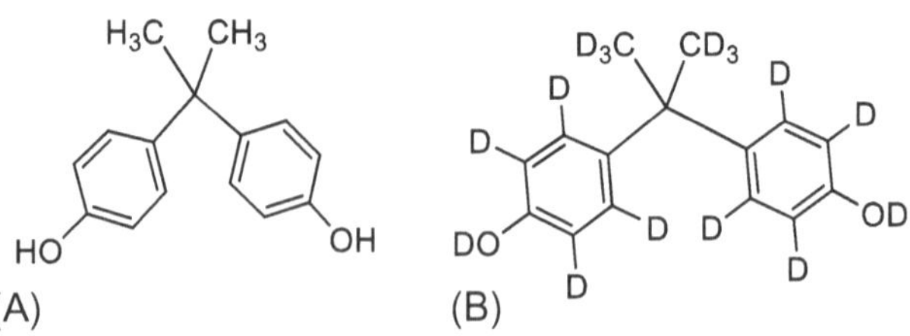

Fig. 19.7.1 (A) Bisphenol A (BPA). (B) BPA-d_{16}. D = deuterium.

5 h and quickly cleared within 24 h. This means the ingested BPA is metabolized rapidly by the liver.

In conclusion, compared to dietary BPA exposure, dermal absorption of BPA leads to longer and more dangerous exposure.

19.8 NEW RADIOLABELED COMPOUNDS FOR PET ^{11}C IMAGING AGENT

19.8.1 Introduction

Positron emission tomography (PET) is a sensitive molecular imaging technique that is extremely important for biomedical research and for drug discovery and development. The value of PET highly depends on the availability of biochemically specific radiotracers. The radiolabel tracers are usually one of the short-lived cyclotron-produced positron-emitters, such as carbon-11 ($t_{1/2}=20$ min) or fluorine-18 ($t_{1/2}=110$ min).

The label position is often critical for avoiding troublesome radiometabolites that may destroy to quantify radiotracer interaction with the imaging target (Pike, 2016). Therefore, in many cases it is preferable to label in one part of the structure rather than in another.

Another important consideration is the radioactivity [Bq] per total mass of tracer [mol]. For low-density imaging targets, such as enzymes, transporters, and receptors, the radiotracer molar activity needs to be very high.

19.8.2 [^{11}C]Fluoroform Labeling

Haskali and Pike (2017) have described the synthesis of [^{11}C]fluoroform based on CoF_3-mediated fluorination of cyclotron-produced [^{11}C]methane. This process is very efficient and reliably repeatable. [^{11}C]Fluoroform is very useful for labeling small molecules in very high molar activity (>200 GBq mmol^{-1}), far exceeding that possible by using [^{18}F]fluoroform. Therefore, [^{11}C]fluoroform represents a major breakthrough for labeling PET tracers in trifluoromethyl groups at high molar activity.

In Fig. 19.8.1A, a total synthesis route of [^{11}C]-labeled antidepressant fluoxetine (Prozac) is shown. The total procedure takes <20 min.

19.8.3 Direct ^{11}CN Labeling of Peptides

Zhao et al. (2017) have developed a practical procedure for ^{11}CN-labeling of unprotected peptides. The method is highly chemoselective for cysteine, and the radiolabeled products were synthesized and purified in <15 min.

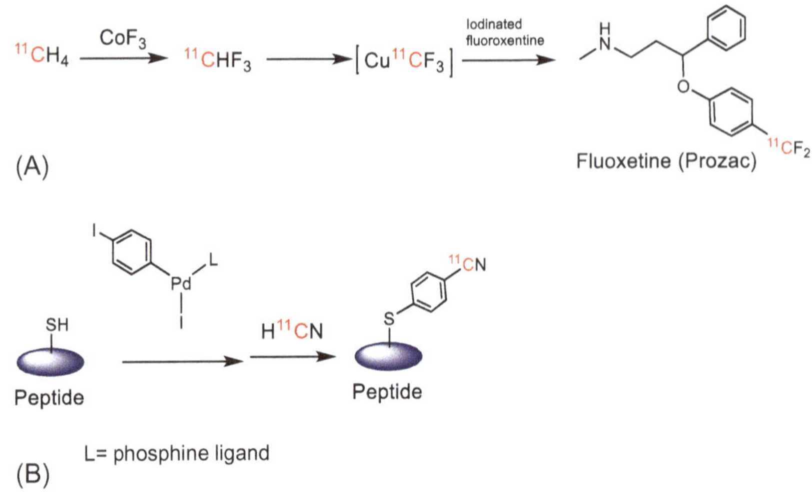

Fig. 19.8.1 New radiolabeled compounds for PET imaging. (A) A total synthesis route of [^{11}C]-labeled fluoxetine (Prozac). (B) A schematic diagram of ^{11}C-labeling of a peptide.

Being appropriate for biomedical applications, the method can be applied on an extremely small scale (20 nmol) with a high radiochemical yield. The success of the protocol came out from the use of a Pd-reagent based on a dihaloarene, which enables direct "nucleophile – nucleophile" coupling of the peptide and [^{11}C]cyanide by temporal separation of nucleophile addition.

A schematic diagram of a ^{11}C-labeling of a peptide is shown in Fig. 19.8.1B.

REFERENCES

Arroyo-Currás, N., Somerson, J., Vieirad, P.A., Ploensee, K.L., Kippine, T.E., et al., 2017. Real-time measurement of small molecules directly in awake, ambulatory animals. PNAS 114, 645–650.

Blackstock, L.K.J., Wang, W., Vemula, S., Jaeger, B.T., Li, X.-F., 2017. Sweetened swimming pools and hot tubs. Environ. Sci. Technol. Lett. 4, 149–153.

Buerge, I.J., Buser, H.R., Kahle, M., Müller, M.D., Poiger, T., 2009. Ubiquitous occurrence of the artificial sweetener acesulfame in the aquatic environment: an ideal chemical marker of domestic wastewater in groundwater. Environ. Sci. Technol. 43, 4381–4385.

Chavali, R., Gunda, N.S.K., Naicker, S., Mitra, S.K., 2014. Detection of *Escherichia coli* in potable water using personal glucose meters. Anal. Methods 6, 6223–6227.

Cui, X., Das, A., Dhawane, A.N., Sweeney, J., Zhang, X., et al., 2017. Highly specific and rapid glycan based amperometric detection of influenza viruses. Chem. Sci. 8, 3628–3634.

Daiber, E.J., DeMarini, D.M., Ravuri, S.A., Liberatore, H.K., Cuthbertson, A.A., et al., 2016. Progressive increase in disinfection byproducts and mutagenicity from source to tap swimming pool and spa water: impact of human inputs. Environ. Sci. Technol. 50, 6652–6662.

Gougeon, R., Spidel, M., Lee, K., Field, C.J., 2004. Canadian diabetes association national nutrition committee technical review: nonnutritive intense sweeteners in diabetes management. Can. J. Diabetes 28, 385–399.
Gurale, B.P., Dhawane, A.N., Cui, X., Das, A., Zhang, X., Iyer, S.S., 2016. Indirect detection of Glycosidases using amperometry. Anal. Chem. 88, 4248–4253.
Harris, A., Cardone, G., Winkler, D.C., Heymann, J.B., Brecher, M., et al., 2006. Influenza virus pleiomorphy characterized by cryoelectron tomography. Proc. Natl. Acad. Sci. U. S. A. 2006 (103), 19123–19127.
Haskali, M.B., Pike, V.W., 2017. [^{11}C]Fluoroform, a breakthrough for versatile labeling of PET radiotracer Trifluoromethyl groups in high molar activity. Chem. Eur. J. 23, 8156–8160.
Hutchins, A., 2016. Why Did the Rio Olympics Pool Water Turn Green? Maclean's. (August 9).
Lin, H.-Y., Huang, C.-H., Park, J., Pathania, D., Castro, C.M., et al., 2017. Integrated magneto-chemical sensor for OnSite food allergen detection. ACS Nano 11, 10062–10069.
Liu, J., Martin, J.W., 2017. Prolonged exposure to bisphenol a from single dermal contact events. Environ. Sci. Technol. 51, 9940–9949.
Mohapatra, H., Phillips, S.T., 2013. Reagents and assay strategies for quantifying active enzyme analytes using a personal glucose meter. Chem. Commun. 49, 6134–6136.
Nakhleh, M.K., Amal, H., Jeries, R., Broza, Y.Y., Aboud, M., et al., 2017. Diagnosis and classification of 17 diseases from 1404 subjects via pattern analysis of exhaled molecules. ACS Nano 11, 112–125.
Pike, V.W., 2016. Considerations in the development of reversibly binding PET Radioligands for brain imaging. Curr. Med. Chem. 23, 1818–1869.
Riglar, D.T., Giessen, T.W., Baym, M., Kerns, S.J., Niederhuber, M.J., et al., 2017. Engineered bacteria can function in the mammalian gut long-term as live diagnostics of inflammation. Nat. Biotechnol. 7, 653–658.
Renwick, A.G., 1986. The metabolism of intense sweeteners. Xenobiotica 16, 1057–1071.
Schmalz, C., Frimmel, F.H., Zwiener, C., 2011. Trichloramine in swimming pools—formation and mass transfer. Water Res. 45, 2681–2690.
Thickett, K.M., McCoach, J.S., Gerber, J.M., Sadhra, S., Burge, P.S., 2002. Occupational asthma caused by chloramines in indoor swimmingpool air. Eur. Respir. J. 19, 827–832.
Van Stempvoort, D.R., Roy, J.W., Brown, S.J., Bickerton, G., 2011. Artificial sweeteners as potential tracers in groundwater in urban environments. J. Hydrol. 401, 126–133.
Villanueva, C.M., Cantor, K.P., Cordier, S., Jaakkola, J.J.K., King, W.D., Lynch, C.F., et al., 2004. Disinfection byproducts and bladder cancer. Epidemiology 15, 357–367.
Wang, Q., Wang, H., Yang, X., Wang, K., Liu, R., Li, Q., Ou, J., 2015. A sensitive one-step method for quantitative detection of α-amylase in serum and urine using a personal glucose meter. Analyst 140, 1161–1165.
Zhang, T., Gan, Z., Gao, C., Ma, L., Li, Y., et al., 2016. Occurrence of artificial sweeteners in human liver and paired blook and urine samples from adults in Tianjin, China and their implications for human exposure. Environ. Sci. Process. Impacts 18, 1169–1176.
Zhang, X., Dhawane, A.N., Sweeney, J., He, Y., Vasireddi, M., et al., 2015. Electrochemical assay to detect influenza viruses and measure drug susceptibility. Angew. Chem., Int. Ed. Engl. 54, 5929–5932. 2015.
Zhao, W., Lee, H.G., Buchwald, S.L., Hooker, J.M., 2017. Direct ^{11}CN-labeling of unprotected peptides via palladium mediated sequential cross-coupling reactions. J. Am. Chem. Soc. 139, 7152–7155.

CHAPTER TWENTY

Nanomachine-Related Topics

Contents

20.1 MOLECULAR DRILL FOR OPENING CELL MEMBRANES

Beyond the more common chemical delivery strategies, there are several physical techniques to open the lipid bilayers of cellular membranes. These include electric and magnetic fields, temperature, ultrasonic, or light to put compounds into cells, to release molecular species from cells, or to induce selective cell death (apoptosis). More recently, molecular motors and switches that can change their conformation in a controlled way in response to external stimuli have been used to produce mechanical actions on tissue for biomedical applications (Barber et al., 2016).

García-López et al. (2017) demonstrated that molecular machines can drill through cellular lipid bilayers using the molecular-scale nanomechanical action. After the adsorption of the molecular motors on lipid bilayers and activation of the motors using ultraviolet light, holes are drilled in the cell membranes.

García-López et al. designed molecular motors (see Fig. 20.1.1) and complementary experimental protocols that use nanomechanical action. They also showed that, by changing functional addends (R), the motors can target specific cell surface sites. After the in vitro applications, they expect that molecular machines could also be used in vivo, especially as their design progresses to allow two-photon, nearinfrared, and radio-frequency activation.

https://doi.org/10.1016/B978-0-12-817054-0.00020-5
© 2019 Elsevier Inc.
All rights reserved.

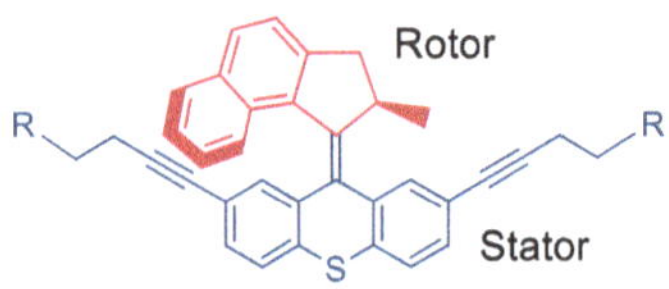

R are functional addends, which can be changed according to cellular targeting.

Fig. 20.1.1 Molecular drill for opening cell membranes.

20.2 ANTIBODY-POWERED DNA NANOMACHINES

Ranallo et al. (2017) reported an antibody-powered DNA-based nanomachine that can reversibly load and release a molecular cargo on binding to a specific antibody. A DNA strand labeled with two antigens can load a nucleic acid strand through a clamp-like triplex-forming mechanism that uses Hoogsteen-interactions (see Fig. 20.2.1). The key is that the DNA clamp is closed by the Hoogsteen interactions (non-Watson-Crick nonpairings), which are weaker than Watson-Crick pairings. However, the DNA clamp opens due to the force when the antibody arrives. The binding of a bivalent macromolecule to the two antigens causes a conformational change that reduces the stability of the triplex complex, with the consequent release of the loaded strand.

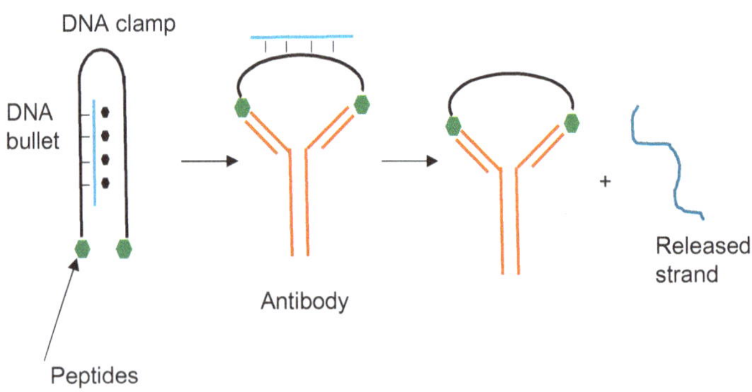

Fig. 20.2.1 The DNA nanomachine. A DNA strand (shown in *black*) labeled with two antigens *(green hexagons)* can load a nucleic acid strand (*blue*) through a clamp-like triplex mechanism which uses Hoogsteen-interactions *(black hexagons)*. The binding of a bivalent macromolecule (an antibody) to the two antigens causes a conformational change that reduces the stability of the triplex complex, with the consequent release of the loaded strand.

20.3 VESICLE ORIGAMI: CUBOID PHOSPHOLIPID VESICLES

When lipids self-asssemble as vesicles, to minimize the surface tension, the vesicles typically become spherical. There are only a few nonspherical systems, but they require templates to scaffold the structure.

However, Neuhaus et al. (2017) reported a 1,2-diaminophospholipid that self-asssembles into a cuboid structure without a scaffold. The bilayer membranes form an exceptionally tight subgel packing, leading to maximization of flat structural elements and minimization of any edges. Wide-angle X-ray scattering measurements suggest that the membrane packs in a herringbone pattern, which is the tightest bilayer packing pattern known. Because of their stiffness, the membranes must be heated above their melting temperature to form vesicles. When the vesicles are subsequently cooled below the melting temperature, they adopt a cuboid shape.

Neuhaus et al. (2017) wanted to use the cubes as drug delivery devices. However, the cuboids themselves are too weak for this purpose because of their very long defect lines along the edges. However, such studies have taught us a lot about the physics needed for the next generation of mechano-responsive drug delivery containers.

20.4 NANOSWIMMERS TO THE BRAIN

Recently, artificial microscopic and nanoscopic self-propelling particles (called nano- or microswimmers) have been created. They are used for transporting target chemicals. One of their most difficult tasks is to deliver drugs within the central nervous system (CNS), because they are guarded by the blood–brain barrier (BBB).

Joseph et al. (2017) noticed that the largest organ concuming glucose and oxygen is the brain, so that there is their large gradient in the body. If the nanoswimmers can detect the gradient, they can concentrate the swimmers into CNS, if possible passing through BBB.

Joseph et al. achieved this by encapsulating glucose oxidase alone or in combination with catalase into nanoscopic and biocompatible asymmetric polymer vesicles (known as polymersomes; see Fig. 20.4.1A and B). They showed that these vesicles self-propel in response to an external gradient of glucose by inducing a slip velocity on their surface, which makes them move in an extremely sensitive way toward higher-concentration regions.

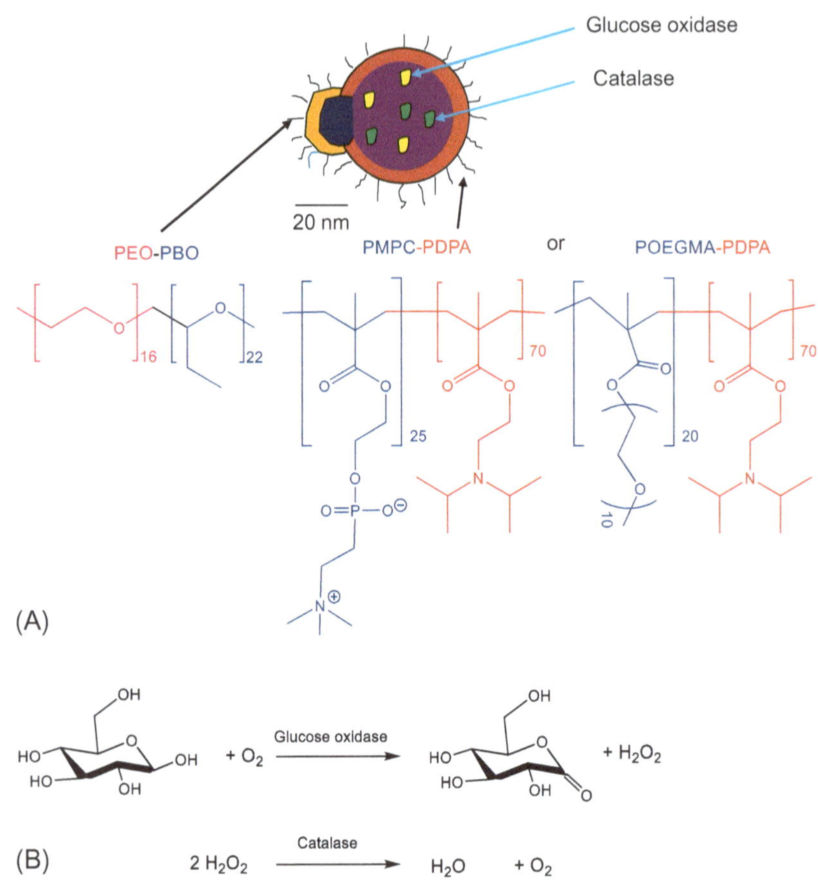

Fig. 20.4.1 Nanno swimmers to the brain. (A) Schematic figure of asymmetric polymersome, and chemical compositions of fibers on each part. (B) Enzymatic reactions of glucose oxidase and catalase.

They finally demonstrated that the chemotactic behavior of these nanoswimmers resulted in a fourfold increase in penetration to the brain compared to nonchemotactic systems.

20.5 HYDROPHOBIC PHAGE-MIMICKING MEMBRANE ACTIVE ANTIMICROBIALS

Jiang et al. (2017) have developed a virus-like polymer that can kill target bacteria, including antibiotic-resistant strains, without harming human cells. In addition, by changing the size and shape of the particle, the target bacteria can be altered.

Jiang et al. (2017) made three polymer nanoparticles: (i) 8 nm wide sphere; (ii) 7 nm wide rod-shaped with 18 nm (short); and (iii) 70 nm (long) length. In the sphere of (i), positively charged, hydrophilic polimer molecular brushes (PMBs) made of poly(4-vinyl-*N*-methylpyridine iodide) (P4MVP) come out from the cores made of β-cyclodextrin (β-CD). For the rod-shaped nanoparticles of (ii) and (iii), PMBs made of P4MVP grow from the rod-like polymers.

Jiang et al. tested nanoparticles to Gram-negative *Escheria Coli*, Gram-positive *Staphylococcus aureus*, and multidrug-resistant bacteria such as *Pseudomonas aeruginosa*. The spherical particles killed 99.9% of both Gram-positive and Gram-negative bacteria and *Pseudomonas aeruginosa*. The longer particles showed lower antibacterial activity, but were more effective against Gram-negative bacteria than Gram-positive ones.

The reason why the nanoparticles are not deadly to human cells is down to the shapes of lipids within cells membranes. The bacterial membranes are rich in phosphoethanolamine lipids with an intrinsic curve, which are easily bent and wrapped by the nanoparticles, resulting in rupture when the nanoparticles gather around. In contrast, human cells are made of flat phosphocholine lipids, forming more rigid membranes.

REFERENCES

Barber, D.M., Schönberger, M., Burgstaller, J., Levitz, J., Weaver, C.D., et al., 2016. Optical control of neuronal activity using a light-operated GIRK channel opener (LOGO). Chem. Sci. 7, 2347–2352.

García-López, V., Chen, F., Nilewski, L.G., Duret, G., Aliyan, A., et al., 2017. Molecular machines open cell membranes. Nature 548, 567–572.

Jiang, Y., Zheng, W., Kuang, L., Ma, H., Liang, H., 2017. Hydrophilic phage-mimicking membrane active antimicrobials reveal nanostructure-dependent activity and selectivity. ACS Infect. Dis. 3, 676–687.

Joseph, A., Contini, C., Cecchin, D., Nyberg, S., Ruiz-Perez, L., et al., 2017. Chemotactic synthetic vesicles: design and applications in blood-brain barrier crossing. Sci. Adv. 3e1700362.

Neuhaus, F., Mueller, D., Tanasescu, R., Balog, S., Ishikawa, T., et al., 2017. Vesicle origami: cuboid phospholipid vesicles formed by template-free self-assembly. Angew. Chem. Int. Ed. 56, 6515–6518.

Ranallo, S., Prévost-Tremblay, C., Idili, A., Vallée-Bélisle, A., Ricci, F., 2017. Antibody-powered nucleic acid release using a DNA-based nanomachine. Nat. Commun. https://doi.org/10.1038/ncomms15150.

CHAPTER TWENTY ONE

Click Chemistry-Related Topics

Contents

21.1 WHAT IS CLICK CHEMISTRY?

The classic click chemistry starts from the Copper-catalyzed Azide-Alkyne Cycloaddition (CuAAC) reaction (Kolb and Sharpless, 2003; see Fig. 21.1.1).

21.2 METHIONINE MODIFICATION IN PROTEINS

For applications such as protein probing, antibody-drug conjugation, and proteomics, cysteine in proteins can be selectively functionalized by numerous acid-base conjugation methods. However, specifical ligation to another sulfur-containing amino acid, methionine, is excluded by its essentially weaker nucleophilicity.

Lin et al. (2017) have developed a chemoselective methionine-bioconjugation method. In this new method, employing redox reactivity, oxaziridine-based reagents enable highly selective, rapid, and robust methionine labeling to be achieved. This reaction is named Redox Activated Chemical Tagging (ReACT), and can be used under biocompatible reaction condition ranges.

An example of the application of the ReACT reaction is shown in Fig. 21.2.1. The protein has a methionine (Fig. 21.2.1A). The oxaziridine compound is added, and the ReACT reaction occurs, functionalizing

Biochemistry for Materials Science
https://doi.org/10.1016/B978-0-12-817054-0.00021-7
© 2019 Elsevier Inc.
All rights reserved.

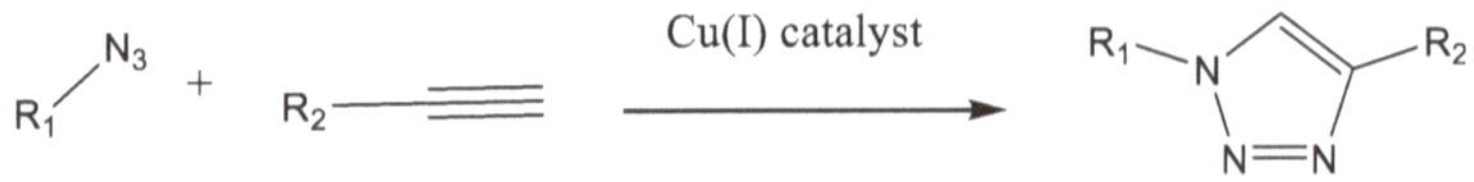

Fig. 21.1.1 CuAAC reaction.

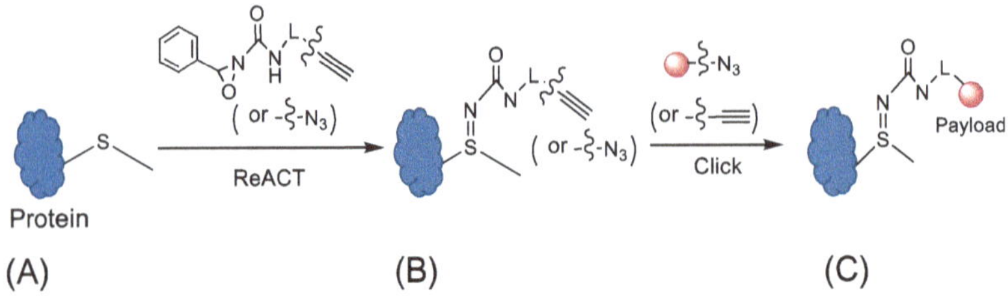

Fig. 21.2.1 The example of the novel methionine conjugation reaction by oxaziridine-based reagents, combined with the Click chemistry for adding a payload. (A) A protein with a methionine. (B) An oxaziridine compound with alkyne or azide functional group is added. (C) The payload with azide or alkyne functional group is added, and the payload is connected to the original protein. ReACT: Redox Activated Chemical Tagging; L: Linker; Click: Click chemistry.

methionine into sulfimide conjugation product (benzaldehyde is a byproduct) (Fig. 21.2.1B). The oxaziridine compound has an alkyne ($—C \equiv C$) or an azide ($—N_3$) functional group via a linker (L). The payload with an azide or an alkyne functional group is then added, and the payload is added to the protein, of which the reaction is called click chemistry (Kolb et al., 2001; see Fig. 21.2.1C). These reactions occur quantitatively, so that precise functionalization of protein can be achieved.

Lin et al. (2017) have also shown the broad utility of this conjugation method to many applications such as synthesis of antibody-drug conjugates and identification of hyperreactive methionine residues.

21.3 A NEW CLICK CHEMISTRY

The foundation of click chemistry (see Section 21.1) was inspired by examination of the central metabolism of life—a process orchestrated by functional oligomers of a few dozen reactive modules, all stitched together via heteroatom links and reversible condensation processes (Kolb et al., 2001; Li et al., 2017).

The discovery of the Cu(I)-catalyzed Azide-Alkyne Cyclo-addition (CuAAC) reaction in 2002 (Tornøe et al., 2002) had a deep impact on the evolution of click chemistry. It demonstrated great usefulness and application in diverse fields such as materials science, bioconjugation, and drug discovery.

Sulfur(VI) fluoride exchange (SuFEx) is a new type of click chemistry that enables the synthesis of covalently linked modules via S^{VI} hubs.

Li et al. (2017) have reported thionyl tetrafluoride (SOF_4) as the first multidimensional SuFEx connector. SOF_4 resides in between the commercially mass-produced gases SF_6 and SO_2F_2, and, like them, is readily synthesized on scale. Under SuFEx catalysis conditions, SOF_4 reliably seeks out primary amino groups [R-NH_2] and becomes permanently anchored via a tetrahedral iminosulfur-(VI) link: R—N=(O=)S(F)$_2$. The selectivity of SO_2F_2 and SOF_4 to aromatic hydroxyl and amino groups is shown in Fig. 21.3.1A.

The difluoride groups R—N=(O=)SF_2 offer two further SuFExable handles, which can be sequentially exchanged to create 3-dimensional covalent departure vectors from the tetrahedral sulfur(VI) hub.

Fig. 21.3.1B shows the multidimensional tetrahedral production starting from thionyl tetrafluoride as an example. First, thionyl tetrafluoride is reacted with primary amino group, forming an iminosulfur product with two fluoride "handles." One or both fluorides can react with aryl alcohols or alkyl amines to produce doubly or triply substituted sulfur along the tetrahedral sulfur hub axes.

Li et al. (2017) have shown the potential applications of this newly arriving click chemistry, such as the syntheses of functional polymers and small-molecule enzyme inhibitors.

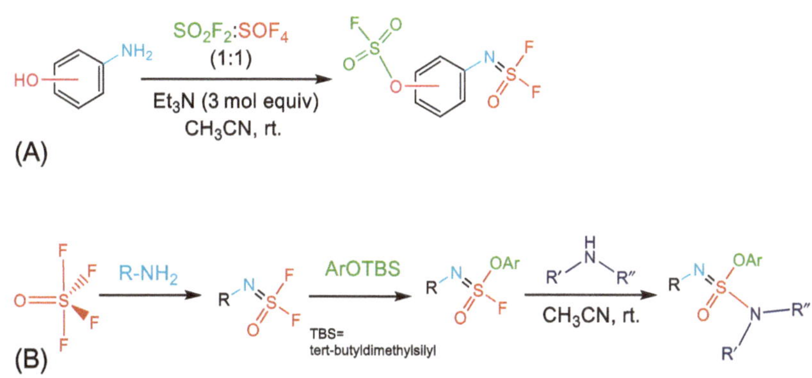

Fig. 21.3.1 (A) Selectivity of SO_2F_2 and SOF_4 to aromatic hydroxyl and amino groups. (B) One example of multidimensional tetrahedral product.

21.4 APPLICATION OF CLICK CHEMISTRY TO SELECTIVE CANCER TARGETING

Many cancer drugs attack not only tumors but also healthy tissues, resulting in harmful side effects. Therefore, doctors wish to target cancer cell more selectively. One such method is antibody-drug conjugates. Antibodies recognize antigens, such as HER2, on some breast cancer cells, allowing antibody-attached drugs to kill these cancer cells selectively. However, antibody-drug conjugates are expensive, and must be administered intravenously.

To overcome these problems, Wang et al. (2017) have developed a strategy named active tissue targeting via anchored click chemistry (ATTACK), which may have advantages over antibody-drug conjugates.

ATTACK is performed in two steps. In the first step, Wang et al. (2017) give tumor-bearing mice an ether-protected sugar which has an azide group (see Fig. 21.4.1A). In more detail, they converted an anomeric acetyl group of tetraacetyl-N-azidoacetylmannosamine (Ac_4ManAz) to a caged ether bond (see Fig. 21.4.1).

Cancer cells metabolize and de-protect the sugar, which is attached to glycoproteins in the cell membrane. Because cancer cells grow rapidly, they overexpress some enzymes, such as histone deacetylase and cathepsin

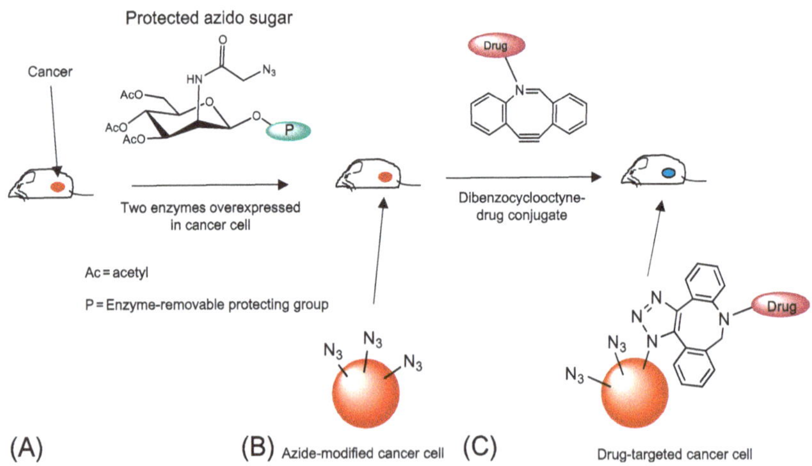

Fig. 21.4.1 Schematic diagram of active tissue targeting via anchored click chemistry (ATTACK). (A) A mouse with cancer. (B) A mouse with azide-modified cancer cells. (C) A mouse with drug-targeted cancer cells.

L-responsive acetylated azidomannosamine, which can selectively cleave and thus overexpress the azido groups on the surface of cancer cells. The cancer cells are more tagged by the azide groups than normal cells (see Fig. 21.4.1B).

In the second step, Wang et al. (2017) give the mice an anticancer drug conjugated to dibenzocyclooctyne (DBCO; see Fig. 21.4.1B). An alkyne group of DBCO reacts with azides by a selective click chemistry reaction, which helps the conjugated cancer drug to fix selectively on target tumors (see Fig. 21.4.1C).

In the mice, ATTACK improved drug-targeting selectivity for treated tumors than those for untreated ones. Compared to the cancer drug doxorubicin alone, a DBCO-doxorubicin conjugate was much less toxic and significantly more effective in treating against LS174T colon cancer, MDA-MB-231 triple-negative breast cancer, and 4T1 metastatic breast cancer in mice.

Future studies on human should be carried out as soon as possible.

21.5 TRANSFECTION OF CELLS BY CLICK CHEMISTRY

In a fundamental biomedical research, the efficient delivery of foreign nucleic acids (transfection) into cells is a critical tool, and a key in several biotechnology industries. There are three main strategies for transfection: (i) reagent; (ii) instrument; and (iii) virus.

Reagent-based methods are used most commonly because of their low cost and ease of use. The method depends on the efficient packaging of a reagent with nucleic acids to form a stable complex, which is subsequently delivered to cells via nonspecific electrostatic targeting. Reagent transfection methods use various polyamine cationic type molecules to condense with negatively charged nucleic acids into a highly positively charged complex. This is subsequently delivered to negatively charged cells in culture for association, internalization, release, and expression.

Although this looks like a straightforward procedure, many problems exist, such as low efficiency, sorting between viable transfected and non-transfected cells, and limited scope of transfectable cell types.

O'Brien et al. (2017) have developed a new strategy (SnapFect; see Fig. 21.5.1) for nucleic acid transfection to cells that does not use electrostatic interactions, but uses an integrated method combining bioorthogonal liposome fusion, click chemistry, and cell surface engineering.

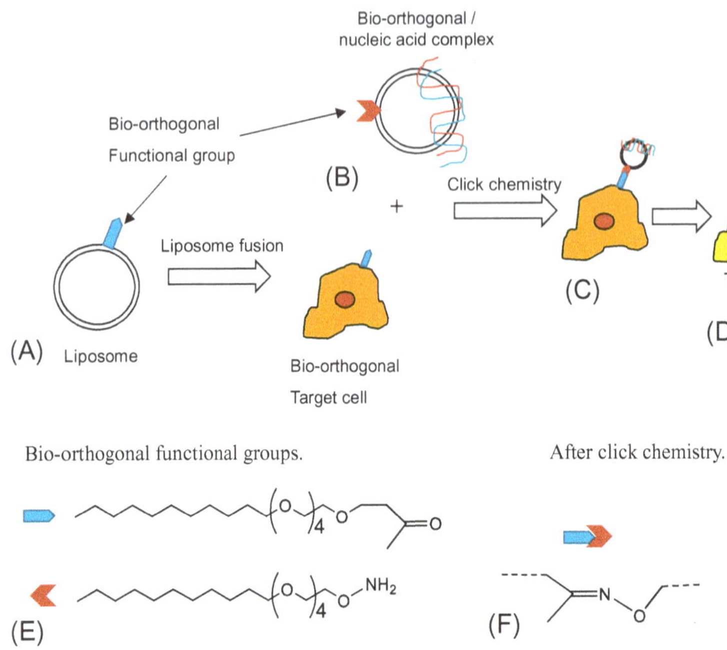

Fig. 21.5.1 Schematic diagram of "SnapFect." (A) A liposome containing a bio-orthogonal lipid-like molecule. (B) Down: The orthogonal molecule is delivered to a cell surface by liposome fusion. Up: A complementary bio-orthogonal nucleic acid/liposome complex is formed. (C) The nucleic acid/liposome complex undergoes a fast click reaction at the cell surface. (D) The nucleic acid/liposome complex is endocytosed, and the nucleic acid is translated. (E) The bio-orthogonal functional groups. (F) The binding after the click chemistry.

A liposome containing a bio-orthogonal lipid-like molecule (Fig. 21.5.1A) is delivered to a cell surface via rapid liposome fusion. A complementary bio-orthogonal nucleic acid/liposome complex (Fig. 21.5.1B) is formed and added to the tailored cell. The nucleic acid/liposome complex undergoes a fast click reaction at the cell surface (Fig. 21.5.1C). The nucleic acid/liposome complex is then endocytosed, and the nucleic acid is translated (Fig. 21.5.1D). The bio-orthogonal functional groups and the binding after the click chemistry are shown in Fig. 21.5.1E and F, respectively.

This new strategy requires minimal time, steps, and reagents, and leads to superior transfection results for a broad range of cell types. The transfection is efficient with high cell viability and does not require a postsorting step to separate transfected from nontransfected cells in the cell population.

O'Brien et al. (2017) showed for the first time a precise transfection strategy where a single cell type in a coculture was target transfected via the bioorthogonal click chemistry.

REFERENCES

Kolb, H.C., Finn, M.G., Sharpless, K.B., 2001. Click chemistry: diverse chemical function from a few good reactions. Angew. Chem. Int. Ed. 40, 2004–2021.

Kolb, H.C., Sharpless, B.K., 2003. The growing impact of click chemistry on drug discovery. Drug Discov. Today 8, 1128–1137.

Li, S., Wu, P., Moses, J.E., Sharpless, K.B., 2017. Multidimensional SuFEx click chemistry: sequential sulfur(VI) fluoride exchange connections of diverse modules launched from An SOF_4 Hub. Angew. Chem. Int. Ed. 56, 2903–2908.

Lin, S., Yang, X., Jia, S., Weeks, A.M., Hornsby, M., et al., 2017. Redox-based reagents for chemoselective methionine bioconjugation. Science 355, 597–602.

O'Brien, P.J., Elahipanah, S., Rogozhnikov, D., Yousaf, M.N., 2017. Bio-orthogonal mediated nucleic acid transfection of cells via cell surface engineering. ACS Cent. Sci. 3, 489–500.

Tornøe, C.W., Christensen, C., Meldal, M., 2002. Peptidotriazoles on solid phase: [1,2,3]-triazoles by regiospecific copper(I)-catalyzed 1,3-dipolar cycloadditions of terminal alkynes to azides. J. Org. Chem. 67, 3057–3064.

Wang, H., Wang, R., Cai, K., He, H., Liu, Y., et al., 2017. Selective in vivo metabolic cell-labeling-mediated cancer targeting. Nat. Chem. Biol. 13, 415–424.

CHAPTER TWENTY TWO

Fluorine-Related Topics

Contents

22.1 A MILD WAY OF CONVERTING PHENOL TO ARYL FLUORIDES

Aryl and heteroaryl fluorides are components in numerous pharmaceuticals and agrochemicals (Purser et al., 2008). Fluorine has unique properties such as high electronegativity, small size, and inertness of C—F bonds. These properties can affect biological activity by modulating lipophilicity, binding affinity, and metabolic stability. However, in spite of the importance of (hetero)aryl fluorides in medicinal chemistry, no mild, general, and selective synthetic methods for the construction of $C(sp^2)$—F bonds have been established.

Schimler et al. (2017) have discovered a novel method for the reaction of phenols with sulfuryl fluoride (SO_2F_2) to give aryl fluorosulfonate (ArOFs) intermediates, which react with tetramethylammonium fluoride ($(CH_3)_4NF$) to give aryl fluorides (see Fig. 22.1.1). All the reactions occur in one pot and under mild conditions (often at room temperature), and the reagents SO_2F_2 and $(CH_3)_4NF$ are both inexpensive.

Biochemistry for Materials Science
https://doi.org/10.1016/B978-0-12-817054-0.00022-9

© 2019 Elsevier Inc.
All rights reserved.

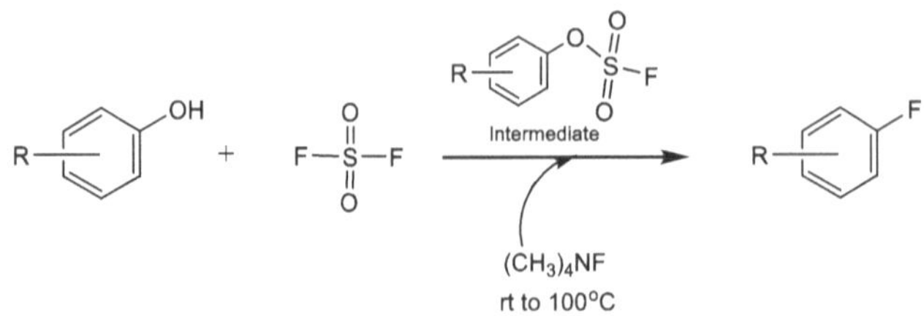

Fig. 22.1.1 Reaction converting phenol to aryl fluorides.

This method will have extensive applications in syntheses of pharmaceutical and agrochemical compounds.

22.2 NEW METHOD FOR SYNTHESIZING CHIRAL FLUORINATED ARENES

22.2.1 Introduction

Trifluoromethyl-substituted cyclopropanes play important roles in medicinal chemistry, because they have the rigidity of three-membered rings with the valuable features of fluorinated substituents. These structural motifs form key building blocks for several drugs.

Many methods have been investigated to add these structures, but many methods use starting materials with the trifluoromethyl group. The most direct method to make such compounds is the addition of trifluoromethylcarbene to an olefin.

Early studies used trifluoromethylcarbene formed by photolytic decomposition of 2-diazo-1,1,1-trifluorethane (CF_3CHN_2; DTE), resulting in a mixture of products. More useful strategies for olefin trifluoromethylcyclopropanation with this reagent have only recently become available by using transition metal catalysis. However, only a few studies have controlled enantioselective variants in the transformations.

In their first report, Simmoneaux and co-workers described the use of chiral metalloporphyrins for the asymmetric cyclopropanation of styrenes in the presence of 2-diazo-1,1,1-trifluoroethane (DTE), but only moderate enantioselectivity (30%–79% ee) was achieved (Le Maux et al., 2006). Higher selectivity was more recently achieved in similar reactions using Co(III)-salen complexes by Carreira and co-workers (Morandi et al., 2011).

Despite these advances, the development of higher stereoselective strategies for olefin cyclopropanation with 2-diazo1,1,1-trifluoroethane as the carbene donor remains a challenge.

22.2.2 Asymmetric Synthesis of Trifluoromethyl-Substituted Cyclopropanes

Tinoco et al. (2017) have reported a simple and efficient biocatalytic approach for promoting these transformations; the approach uses engineered myoglobin variants in combination with a compartmentalized reaction setup. This strategy can be applied to the conversion of a broad range of vinylarene substrates. The synthesized trifluoromethyl-substituted cyclopropanes have high diastereo- and enantioselectivity as well as stereocomplementary selectivity.

These transformations were realized using a two-compartment setup in which ex situ generated gaseous CF_3CHN_2 (DTE) is processed by engineered myoglobin catalysts, Mb(H64V,V68A)*, expressed in bacterial cells (see Fig. 22.2.1A for the experiment setup). This approach was successfully applied to give a variety of trans-1-trifluoromethyl-2-arylcyclopropanes

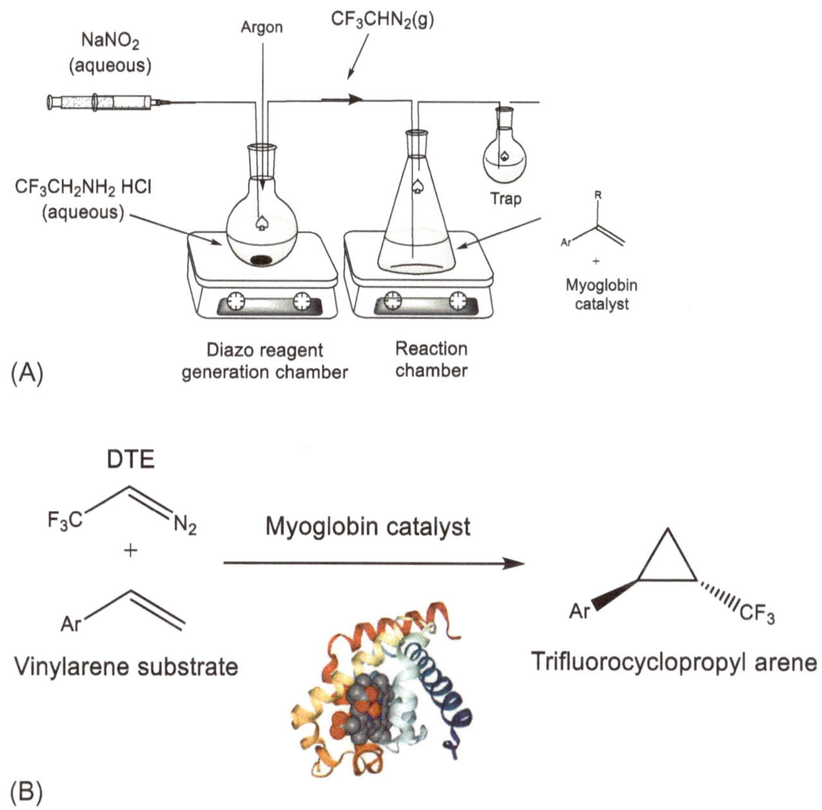

Fig. 22.2.1 (A) The experiment setup of Tinoco et al. (2017). (B) Schematic diagram for the reaction. Ar = arene. DTE = 2-diazo-1,1,1-trifluoroethane.

(see Fig. 22.2.1B) in high yields (61%–99%) and excellent diastereo and enantioselectivity (97%–99.9% de and ee). Furthermore, mirror-image forms of these products could be obtained using myoglobin variants (e.g., Mb(H64V,V68S)) featuring stereodivergent selectivity.

These reactions provide a convenient and effective biocatalytic route to the stereoselective synthesis of key fluorinated building blocks of high value for medicinal chemistry and drug discovery.

This work widens the range of carbene-mediated transformations accessible via metalloprotein catalysts and introduces a potentially general strategy for exploiting gaseous and/or hard-to handle carbene donor reagents in biocatalytic carbene transfer reactions.

(*) Mb(H64V,V68A) means that the 64th and the 68th amino acids of histidine (H) and valine (V) in myoglobin are exchanged by bioengineering into valine (V) by and alanine (A), respectively. The alphabet of amino acids is shown in Table 2.1.1. Details of the bioengineering are shown in Chapter Three.

22.3 CONTINUOUS FLOW SYNTHESIS OF CIPROFLOXACIN HYDROCHLORIDE

Ciprofloxacin (see Fig. 22.3.1) belongs to the family of fluoroquinolone antibiotics. It is used to treat a number of types of bacterial infections, and is on the World Health Organization List of Essential Medicines. Bayer AG developed a seven-step synthesis in the 1980s, with an overall yield

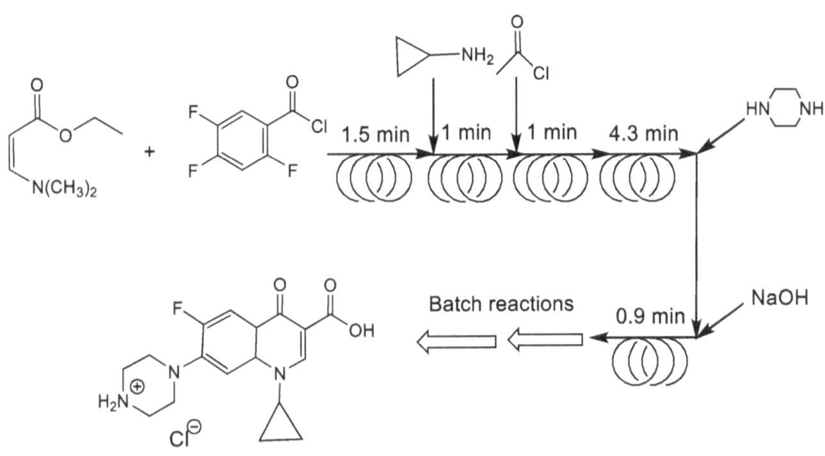

Fig. 22.3.1 Flow scheme of continuous total synthesis of ciprofloxacin hydrochloride.

of 49% and >24 h of reaction time. Later, a similar sequence of reactions were performed with a slightly increased yield of 57% but with >100 h reaction time.

Lin et al. (2017) developed a rapid total synthesis of ciprofloxacin hydrochloride by continuous flow methods. Within a total residence time of 9 min, the sodium salt of ciprofloxacin was synthesized from simple building blocks via a linear sequence of six chemical reactions in five flow reactors (Fig. 22.3.1). Sequential offline acidifications and filtrations gave ciprofloxacin and ciprofloxacin hydrochloride. The overall yield of the eight-step sequence was 60%. No separation of intermediates was required throughout the synthesis when a single acylation reaction was applied to remove the main byproduct, dimethylamine.

22.4 CONVERTING BENZENE RING TO CYCLOHEXENES BY THE COMPLEX TpW(NO)(PMe$_3$)(η^2-PhCF$_3$)

Wilson et al. (2017) explored the effects of an electron-withdrawing group on the organic chemistry of an η^2-bound benzene ring using the complex TpW(NO)(PMe$_3$)(η^2-PhCF$_3$) (Me = methyl, Tp = tris(pyrazolyl) borate; Fig. 22.4.1A).

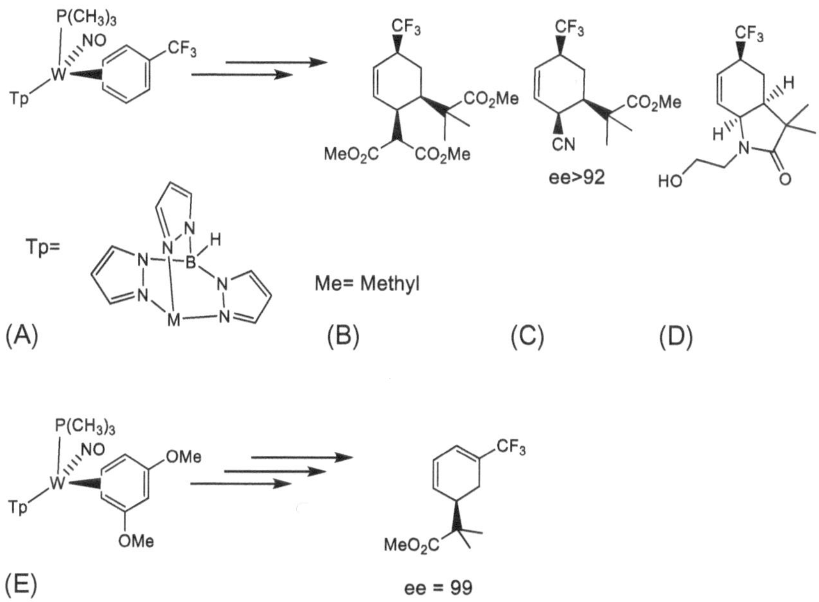

Fig. 22.4.1 Converting benzene ring to cyclohexenes by the complex TpW(NO)(PMe$_3$)(η^2-PhCF$_3$). (A) The starting material. (B)–(D) Examples of stereoselective reactions. (E) Enantio-enriched trifluorotoluene complex.

They found that this trifluorotoluene complex first undergoes a highly regio- and stereoselective 1,2-addition reaction. The resulting 1,3-diene complexes then undergo the second 1,4-addition, again with a high degree of regio- and stereocontrol (see Figs. 22.4.1B–D).

The ability to synthesize enantio-enriched organics was demonstrated for a diene and a trisubstituted cyclohexene. Substitution from an enantio-enriched η^2-dimethoxybenzene complex in neat trifluorotoluene yielded enantio-enriched trifluorotoluene complex, which was elaborated into cyclohexadienes (Fig. 22.4.1E) and cyclohexenes (Fig. 22.4.1C) with ee's of 92% and 99%, respectively.

REFERENCES

Le Maux, P., Le Maux, P., Juillard, S., Simonneaux, G., 2006. Asymmetric synthesis of trifluoromethylphenyl cyclopropanes catalyzed by chiral metalloporphyrins. Synthesis 2006, 1701–1704.

Lin, H., Dai, C., Jamison, T.F., Jensen, K.F., 2017. A rapid Total synthesis of ciprofloxacin hydrochloride in continuous flow. Angew. Chem. Int. Ed. 56, 8870–8873.

Morandi, B., Mariampillai, B., Carreira, E.M., 2011. Enantioselective cobalt-catalyzed preparation of Trifluoromethyl-substituted cyclopropanes. Angew. Chem. Int. Ed. 50, 1101–1104.

Purser, S., Moore, P.R., Swallow, S., Gouverneur, V., 2008. Fluorine in medicinal chemistry. Chem. Soc. Rev. 37, 320–330.

Schimler, S.D., Cismesia, M.A., Hanley, P.S., Froese, R.D.J., Jansma, M.J., et al., 2017. Nucleophilic deoxyfluorination of phenols via aryl fluorosulfonate intermediates. J. Am. Chem. Soc. 139, 1452–1455.

Tinoco, A., Steck, V., Tyagi, V., Fasan, R., 2017. Highly diastereo- and enantioselective synthesis of trifluoromethyl-substituted cyclopropanes via myoglobin-catalyzed transfer of trifluoromethylcarbene. J. Am. Chem. Soc. 139, 5293–5296.

Wilson, K.B., Myers, J.T., Nedzbala, H.S., Combee, L.A., Sabat, M., et al., 2017. Sequential tandem addition to a tungsten – trifluorotoluene complex: a versatile method for the preparation of highly functionalized Trifluoromethylated cyclohexenes. J. Am. Chem. Soc. 139, 11401–11412.

PART FIVE

Environmental Sciences, Origin and Evolution of Life-Related Molecules and Water on Other Planets, Moons and Dwarf Planets

CHAPTER TWENTY THREE

Topics of Environmental Sciences

Contents

23.1 FATE OF NATURAL ISOPRENE EMISSIONS

About 500 Tg of 2-methyl-1,3-butadiene (isoprene) is produced by deciduous trees each year. Isoprene oxidation in the atmosphere starts by addition of hydroxyl radicals (OH) to C_1 or C_4 (see Fig. 23.1.1A) in a ratio of 0.57 to produce two sets of distinct allylic radicals (see Fig. 23.1.1B and C). Oxygen (O_2) adds to these allylic radicals either δ (*Z* or *E* depending on the radicals of cis or trans) or β to the OH group, forming six distinct peroxy radical isomers. The allylic radicals then undergo bimolecular reactions with NO, O_2, or HO_2.

Teng et al. (2017) used isomer-resolved measurements of the reaction products of the peroxy radicals to diagnose this complex chemistry by laboratory studies using an environmental chamber. They found that the ratio of δ to β hydroxy peroxy radicals depends on their bimolecular lifetime ($\tau_{bimolecular}$). At $\tau_{bimolecular} \approx 0.1$ s, a transition occurs to thermodynamically controlled distribution at 297 K. As $\tau_{bimolecular}$ of nature is >10 s, the distribution of isoprene hydroxy peroxy radicals will be controlled by the stability of the peroxy radicals. The β hydroxy peroxy radical isomers comprise ~95% of the radicals, which are not initially formed in such large proportion. However, previous studies used the thermodynamically controlled distribution.

https://doi.org/10.1016/B978-0-12-817054-0.00023-0
© 2019 Elsevier Inc.
All rights reserved.

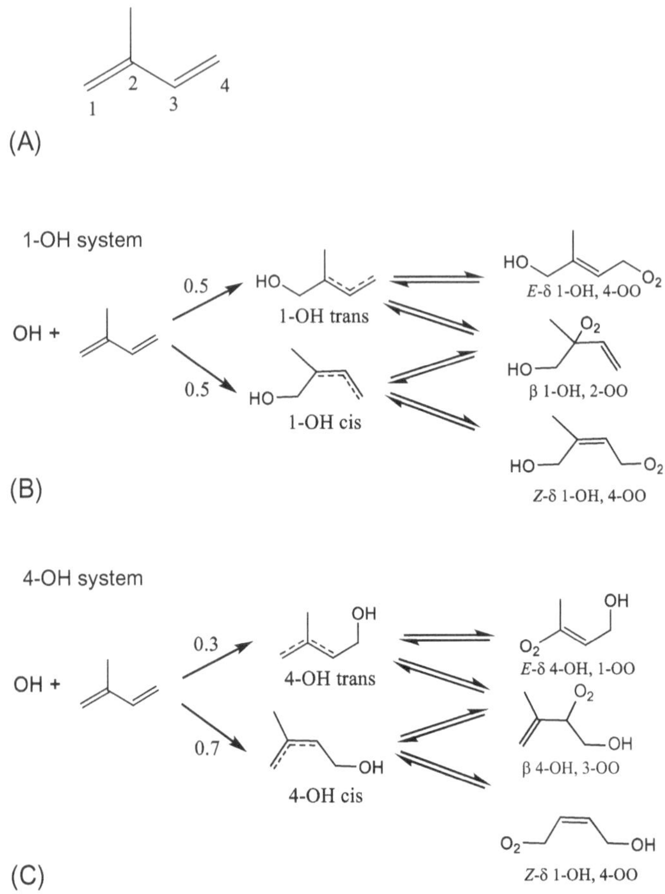

Fig. 23.1.1 (A) 2-Methyl-1,3-butadiene (isoprene). (B) and (C). Two separate systems of peroxy radicals by OH addition to isoprene. 1-OH system (B) and 4-OH system (C).

23.2 NITROGEN FIXING IS AFFECTED BY OCEAN ACIDIFICATION

An increase of carbon dioxide (CO_2) in the surface ocean decreases the seawater pH. As result, nitrogen fixation, where nitrogen in air is converted into NH_3, is affected because >50% of the nitrogen fixation is done by the cyanobacteria *Trichodesmium*. However, studies on how the CO_2 increase in seawater affects nitrogen fixing have given contradicting results.

The new results by Hong et al. (2017) indicated that higher CO_2 levels increase *Trichodesmium* growth, resulting in higher nitrogen fixation. However, lower pH leads to lower growth of the bacterium, which is overridden on the net growth. Furthermore, the negative effect is amplified by the lack of nutrients, such as iron.

23.3 NITRIC OXIDE (NO) PLAYS AN IMPORTANT ROLE IN THE GLOBAL ECO-SYSTEM

Substantial amounts of nitric oxide (NO) and nitrous oxide (N_2O), which cause harmful environmental effects for large-scale agriculture, are emitted by ammonia-oxidizing bacteria (AOB). NO produces the ground-level ozone and acid rain. N_2O is an ozone-depleting greenhouse gas with a global warming potential ~300 times larger than that of CO_2.

The presently accepted model for AOB metabolism of *Nitrosomonas europaea* contains NH_3 oxidation to nitrite (NO_2^-) via a single obligate intermediate, hydroxylamine (NH_2OH).

$$NH_3 + O_2 + 2e^- + 2H^+ \rightarrow NH_2OH + H_2O \qquad (23.3.1)$$

In this model, the multiheme enzyme hydroxylamine oxidoreductase (HAO) catalyzes the four-electron oxidation of NH_2OH to NO_2.

$$NH_2OH + H_2O \rightarrow NO_2^- + 4e^- + 5H^+ \qquad (23.3.2)$$

However, Carantoa and Lancastera (2017) demonstrated evidence that HAO oxidizes NH_2OH, which is made by (23.3.3), by only three electrons to NO under both anaerobic and aerobic conditions.

$$NH_2OH \rightarrow NO + 3H^+ + 3e^- \qquad (23.3.3)$$

NO_2^- observed in HAO activity analysis is a nonenzymatic product resulting from the oxidation of NO by O_2 under aerobic conditions.

$$NO + 1/2\,O_2 + e^- \rightarrow NO_2^- \qquad (23.3.4)$$

The study of Carantoa and Lancastera (2017) suggested that aerobic NH_3 oxidation by AOB occurs via two obligate intermediates, NH_2OH and NO, necessitating a mediator of the third enzymatic step to NO_2^- (see Fig. 23.3.1).

$$NO + H_2O \rightarrow NO_2^- + 2H^+ + e^- \qquad (23.3.5)$$

In their model, the following are side reactions.

$$NO + H^+ + e^- \rightarrow 1/2\,N_2O + 1/2\,H_2O \qquad (23.3.6)$$

$$NO + O_2 + e^- \rightarrow NO_3^- \qquad (23.3.7)$$

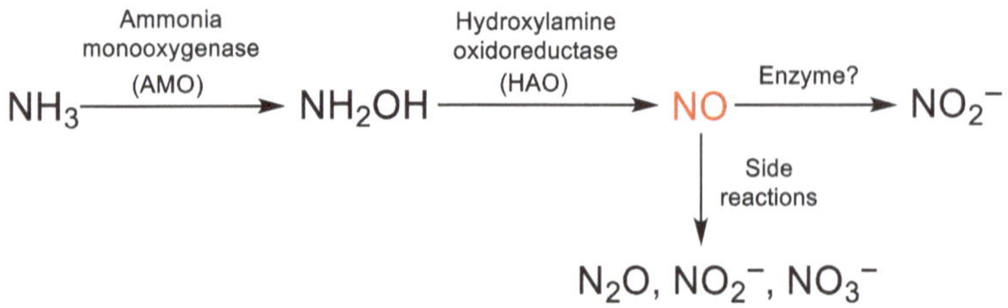

Fig. 23.3.1 The new proposed route to oxidize NH_3 to NO_2^- by Carantoa and Lancastera (2017).

23.4 METHANE FORMATION FROM CO_2 BY AN IRON CATALYST

Converting CO_2 into fuel or raw chemical materials could reduce fossil fuel consumption and climate-changing CO_2 emissions. One strategy aims for electrochemical conversions powered by electricity from renewable sources, but photochemical approaches driven by sunlight are also possible.

A considerable challenge in both approaches is the development of efficient and selective catalysts, ideally based on cheap and Earth-abundant elements rather than expensive precious metals. Of the molecular photo- and electrocatalysts reported, only a few catalysts are stable and selective for CO_2 reduction; moreover, these catalysts produce primarily CO or HCOOH, and catalysts capable of generating even low to moderate yields of highly reduced hydrocarbons remain rare.

Rao et al. (2017) developed an iron tetraphenylporphyrin complex functionalized with trimethylammonio groups (see Fig. 23.4.1A), which is the most efficient and selective molecular electro-catalyst for converting CO_2 to CO known. The iron tetraphenylporphyrin complex could catalyze the eight-electron reduction of CO_2 to methane upon visible light irradiation at ambient temperature and pressure. Rao et al. (2017) also found that the catalytic system (see Fig. 23.4.1C), operated in an acetonitrile solution containing a photosensitizer and sacrificial electron donor, remaining stable over several days. CO is the main product of the direct CO_2 photoreduction reaction, but a two-pot procedure that first reduces CO_2 and then reduces CO generates methane with a selectivity of up to 82% and a quantum yield (light-to-product efficiency) of 0.18%.

Rao et al. (2017) thought that the operating principles of their system might aid the development of other molecular catalysts for the production of solar fuels from CO_2 under mild conditions.

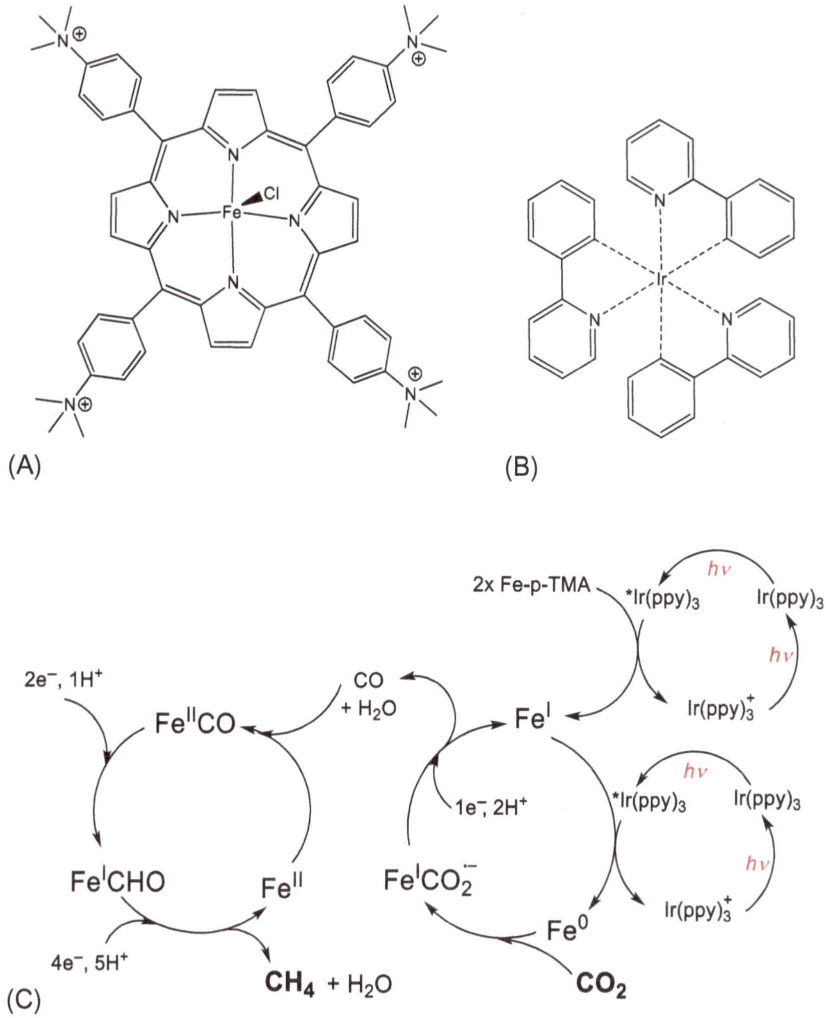

Fig. 23.4.1 (A) Fe-p-TMA. (B) $Ir(ppy)_3$. (C) The proposed mechanism for CO_2 reduction to CH_4 by catalyst (A).

REFERENCES

Carantoa, J.D., Lancastera, K.M., 2017. Nitric oxide is an obligate bacterial nitrification intermediate produced by hydroxylamine oxidoreductase. PNAS 114, 8217–8222.

Hong, H., Shen, R., Zhang, F., Wen, Z., Chang, S., et al., 2017. The complex effects of ocean acidification on the prominent N_2-fixing cyanobacterium *Trichodesmium*. Science 356, 527–531.

Rao, H., Schmidt, L.C., Bonin, J., Robert, M., 2017. Visible-light-driven methane formation from CO_2 with a molecular iron catalyst. Nature 548, 74–77.

Teng, A.P., Crounse, J.D., Wennberg, P.O., 2017. Isoprene peroxy radical dynamics. J. Am. Chem. Soc. 139, 5367–5377.

FURTHER READING

Asara, J.M., Schweitzer, M.H., Freimark, L.M., Phillips, M., Cantley, L.C., 2007. Protein sequences from mastodon and *Tyrannosaurus rex* revealed by mass spectrometry. Science 316, 280.

CHAPTER TWENTY FOUR

Origin and Evolution of Life-Related Molecules

Contents

24.1 INTRODUCTION TO EXPERIMENTS ON ORIGIN AND EVOLUTION OF LIFE-RELATED MOLECULES

In this chapter, experimental developments on origin and evolution of life-related molecules are discussed. These life-related molecules may be divided into two groups. One group is the molecules made of C, H, and O, such as aldoses, ketoses, and sugars including ribose (see Fig. 24.1.1A and B). These molecules are known as CHOs. Another group is made of C, H, O, and N, such as formamide, nucleobases, and amino acids. These molecules are called CHONs. Both CHOs and CHONs are important molecules for life. Inorganic materials are not included in this chapter.

Biochemistry for Materials Science
https://doi.org/10.1016/B978-0-12-817054-0.00024-2

© 2019 Elsevier Inc.
All rights reserved.

Aldoses

		ppm (w/w)		ppm (w/w)
C-2	Ethylene glycol	550	C-5	
	Glycolaldehyde	2390	Ribitol	560
	Glycolic acid	6330	Ribose	260
C-3	Glycerol	2860	Ribonic acid	82
	Glyceraldehyde	302	Arabitol	1150
	Glyceric acid	2440	Arabinose	200
C-4	Erythritol	5070	Arabinoic acid	165
	Erythrose	*	Xylitol	630
	Erythronic acid	960	Xylose	240
	Threitol	7200	Xylonic acid	67
	Threose	**	Arabitol	1150
	Threonic acid	840	Lyxose	145
			Lyxonic acid	140

(A)

Ketoses

		ppm (w/w)
C-3	Dihydroxyacetone	540
C-4	Erythrulose	37
C-5	Ribulose	2010
C-5	Xylulose	470

(B)

Fig. 24.1.1 Structures, names and abundances of linear and branched. (A) Aldoses, (B) ketoses detected in the experiment products of Meinert et al. (2016). The concentration is ppm (weight/weight). Erythrose (*) and threose (**) were below the quantification limit and detection limit, respectively. Identified C-6 analytes are not shown.

There are several proposals on the origin of CHOs and CHONs on Earth:

(I) Large energy such as electrical spark or ultraviolet light activated organic synthesis of CHONs on Earth, as proposed by the classic experiment of Miller (1953).

(II) CHOs and CHONs in comets and asteroids fell to Earth by large heavy bombardment (LHB) without decomposition, as carbonaceous chondrites containing them fell on Earth (Section 24.3).

(III) Strong UV light produced CHOs in space, and fell to Earth (Section 24.4).

(IV) Impact shocks (icy comet-icy comet or icy comet-Earth) in space or on Earth with extremely high energy synthesized CHONs (Sections 24.5–24.7).

In order to prove these hypotheses, various experiments have been performed for synthesizing CHOs and CHONs. In these sections, up-to-date experiments are presented following these categories.

24.2 COUNTERARGUMENTS TO THE COSMIC ORIGIN OF CHOs AND CHONs

There are some counterarguments to the cosmic origin of CHOs and CHONs. Firstly, the amounts of CHONs in IDPs are unclear. Secondly, organic molecules such as CHONs in carbonaceous chondrites and/or interplanetary dust particles (IDPs) may be very fragile.

The experiments of synthesis of CHONs were continued by Oró's group (Oró, 1961). The organic synthesis of CHONs by glow-discharge was continued by Harada and Suzuki (1977). Miyakawa et al. (2002) proposed a cold origin of life.

Along these terrestrial origins of life, impact shock origin of CHOs and CHONs became one of the strong counterarguments. This is the third proposal of Chyba and Sagan (1992), that the synthesizing energy was obtained from the impact shocks of meteorites. The largest difference from the cosmic origin is that the carbons of organic molecules were supplied from the terrestrial atmosphere, because partial pressure of CO_2 was high enough in Hadean and Archaean.

24.3 IMPACT-SHOCK EXPERIMENTS FOR SYNTHESES OF CHOs AND CHONs

The most probable way to synthesize CHOs and CHONs considered today is the impact shock between two icy comets in space or between an icy comet and Earth when they enter the Archaean atmosphere and fall into the Archean Ocean. Three cases were easily imaginable, and were experimentally simulated. The first case is that an icy comet hit another one in space (Meinert et al., 2016). The second case covers both reactions of ice-ice and ice-Earth collisions because only the energy drives the chemical reactions (Ferus et al., 2015). Therefore, the collisional energy as well as the energy from the high energy laser are important. The final experiment simulated the case that an icy comet fell on the early Ocean of Earth (Furukawa et al., 2015), to which the starting materials needs to be paid attention. Hence, Section 24.4 aims to understand the synthesis of CHOs, while Sections 24.5–24.8 target the syntheses of CHONs.

The simulated reaction, starting materials, irradiation medium, experimental condition, target molecules, and their detection methods in each modern experiment are summarized in Table 24.3.1.

24.4 SYNTHESES OF CHOs BY UV IRRADIATION ON INTERSTELLAR ICES

Meinert et al. (2016) published the experiment that CHOs (aldoses and ketoses) were made by ultraviolet irradiation of the interstellar ice condition, composed of water (H_2O), methanol (CH_3OH), and ammonia (NH_3).

The largest difference between Meinert's experiment and other experiments in Table 24.3.1 is that this experiment targeted to synthesize aldoses and ketoses, and finally to produce ribose, which is the backbone of RNA.

In the experiment of Meinert et al. (2016), the samples were prepared after Nuevo et al. (2007). Briefly, the starting material gases (H_2O, $^{13}CH_3OH$, and NH_3 in proportions of 10:3.5:1) were deposited on a MgF_2 substrate cooled at 78 K in 10^{-5} Pa as a thin film of ice. The MgF_2 substrate is transparent from visible to UV light. The gas-deposited substrate was then irradiated by UV light for 142 h, using an H_2 discharge lamp providing essentially Lyman-α photons at 122 nm with a tail including an H_2 recombination line at approximately 160 nm and a continuum down to the visible range. The ratio of ultraviolet photons to deposited molecules was around one.

The substrate was further irradiated at room temperature with right-hand circularly polarized synchrotron radiation (CPSR) at 10.2 eV for 2 h at beamline DESIRS of the SOLEIL (Source Optimisée de Lumière d'Énergie Intermédiaire du Lure in France) synchrotron, which is the French national synchrotron facility. Because Meinert et al. initially intended to induce asymmetric photochemical reactions, no stereochemical effects induced by CPSR could be identified. Photon-molecule interactions predominantly occurred.

The sample was washed with water, derivatized, and analyzed by GCxGC-TOFMS. Briefly, two gas-chromatographic columns are connected for two-dimensional measurement with a reflectron time-of-flight (TOF) mass spectrometer for recording retention times and mass spectra.

The experiment products of the ice samples contained aldoses and ketoses, as shown in Fig. 24.1.1A and B, respectively. Erythrose and threose were below the quantification limit ($<10\sigma$) and detection limit ($<3\sigma$),

Table 24.3.1 Summary of Experimental Conditions for Synthesizing Life-Related Organic Molecules

Researcher	Simulated Reaction	Starting Materials	Irradiation	Experimental Condition	Target Molecules	Detection
Meinert et al. (2016)	Icy grains at molecular cloud stage	Ice Methanol NH_3	Ultraviolet light	10^{-5} Pa, 78 K	CHOs	GCxGC-TOFMS
Martins et al. (2015)	Icy comet/icy comet collision	NH_3 solution Dry ice (CO_2) Methanol	Gas gun	>50GPa	CHONs	GC-MS
Ferus et al. (2015)	Icy comet/icy comet or Earth collision (LHB)	Formamide($HCONH_2$) Olivine chondrite	High energy laser	>4500 K	CHONs	GC-MS
Furukawa et al. (2015)	Icy comet fell into seawater	Fe, Ni (Fe_3O_4), NH_4HCO_3(Mg_2SiO_4) H_2O, N_2 gas	Gas gun		CHONs	UHPLC-MS/MS

respectively. In other words, almost all aldoses and ketoses with carbon number from 2 to 5 were synthesized more than the quantification levels. Meinert et al. detected monosaccharide, ribose, arabinose, xylose, and lyxose, which belong to aldopentoses, and ribuloses and xylulose, which belong to ketopentoses.

24.5 THE EXPERIMENT WHERE ONE ICY COMET HITS ANOTHER IN SPACE

In this experiment by Martins et al. (2015), the case where an icy comet hits another one in space is assumed. Using a gas-gun, life-related molecules of CHONs are synthesized in this condition.

Both carbon dioxide and ammonia are reasonably available as starting materials in the impact environments for prebiotic organic synthesis. Methanol (CH_3OH) is one of the simplest compounds of carbon, and also available in space. Thus, the ice made of mixture of ammonia solution, CO_2, and methanol composed of 9.1:8:1 was made. Two experiments were performed. The ice was impacted twice at 7.15 and 7.00 $km\,s^{-1}$ by a steel projectile. The two pieces (one piece is ~100 g) of ice were used in one collision experiment (one as the control), but <1 mg was at the peak pressure of >50 GPa. The ice piece was recovered, dried, and analyzed by gas chromatography-mass spectrometry (GC-MS). The detection limits for amino acids were ~10 pg.

The impact shock by the ice-ice collision produced several amino acids, including linear and methyl α-amino acids. Linear α-amino acids detected ranged from C_2 to C_5, and included glycine, D- and L-alanine, α-aminobutyric acid (α-ABA), and D- and L-norvaline (see Fig. 24.5.1). Methyl α-amino acids include the nonprotein amino acids isovaline and α-aminoisobutyric acid (α-AIB) (see Fig. 24.5.2; "Non-proteinogenic" means that the amino acid is NOT used in formation of naturally existing proteins.).

A racemic mixture of alanine was detected, with a *D/L* ratio of 0.99 ± 0.05 (target ice sample no. 1) and 0.99 ± 0.02 (target ice sample no. 2). A racemic mixture of norvaline was also detected, with a *D/L* ratio of 0.97 ± 0.04 (target ice sample no. 1) and 0.97 ± 0.02 (target ice sample no. 2). This clearly indicates that there were no contributions (contamination) of terrestrial materials, because all terrestrial samples are made of L-amino acids, resulting in very low *D/L* ratio if the experiments were contaminated.

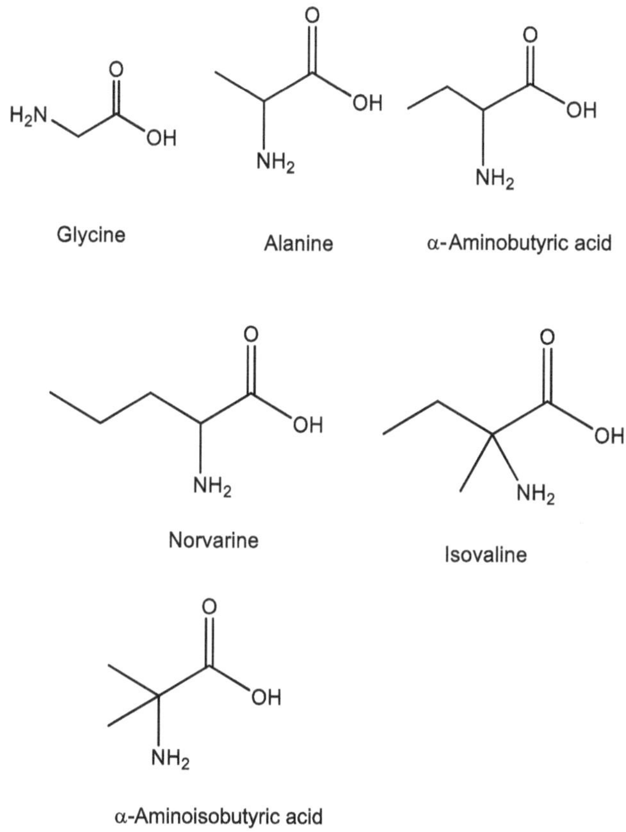

Fig. 24.5.1 Compounds formed by the ice-ice collision experiment.

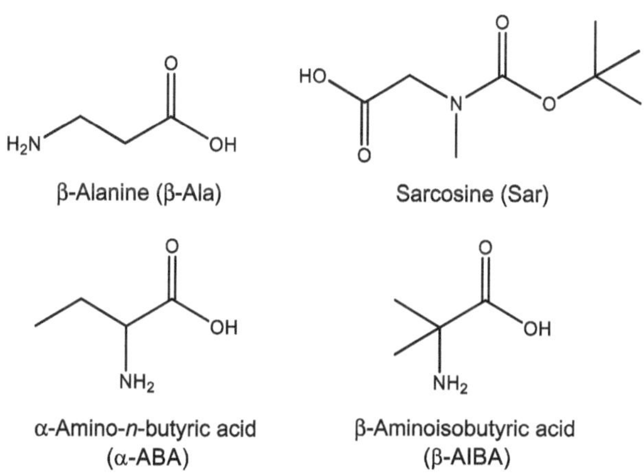

Fig. 24.5.2 Nonproteinogenic amino acids synthesized by the impact experiments.

24.6 THE ICY PLANET AND ICY PLANET COLLISION

A collision between two icy planets is assumed to be the origin of the source of the biogenic molecules or CHONs. To test this assumption and simulate the collision, Ferus et al. (2015) heated the sample by a high-power laser, and analyzed run products by gas chromatography-mass spectrometry (GC-MS).

The plasma formed by the impact of an extraterrestrial body was simulated using the high-power chemical iodine Prague Asterix Laser System (PALS). During the dielectric breakdown in gas (Laser-Induced Dielectric Breakdown or LIDB) generated by a laser pulse of energy of 150J (time interval of ~350ps, wavelength of 1.315 μm, and output density of 10^{14} to $10^{16}\,W\,cm^{-2}$), the outcomes for a high-energy density event occurred. The shock raised the temperature to 4500K, and generated secondary hard radiation (UV and X-ray).

The unstable radicals produced in the formamide dissociation have been identified and quantified using time-resolved discharge emission spectroscopy.

In the LIDB plasma experiments by Ferus et al. (2015), formamide is first decomposed into radicals. Radicals of CN• and NH• are the most abundant species. ("•" indicates a radical, which is a free bond. For example, NH• shows a state that one electron of a lone electron pair of N is removed. Therefore, this nitrogen is very reactive.) The absorption gas-phase spectra showed the presence of HCN, CO, NH_3, CO_2, CH_3OH, and N_2O.

Ferus et al. (2014) reported that total yields of the radicals of CN•, NH• and stable CO were 55%, 4%, and 41% in the impact plasma, respectively. Due to the very rigid structure of CN• from the strong triple bond, this molecule is able to adopt a series of excited electron configurations. These properties make this transient species an ideal reactant in the high-energy plasma environment. Therefore, a CN• radical was also discovered in interstellar space in the envelopes of giant stars. It was known that the stepwise addition of CN• radicals to formamide with atomic H can give rise to the formation of 2,3-diaminomaleonitrile (DAMN) in a highly exergonic reaction. Thus, DAMN (see Fig. 24.6.1) is generally considered to be the common precursor of all nucleobases.

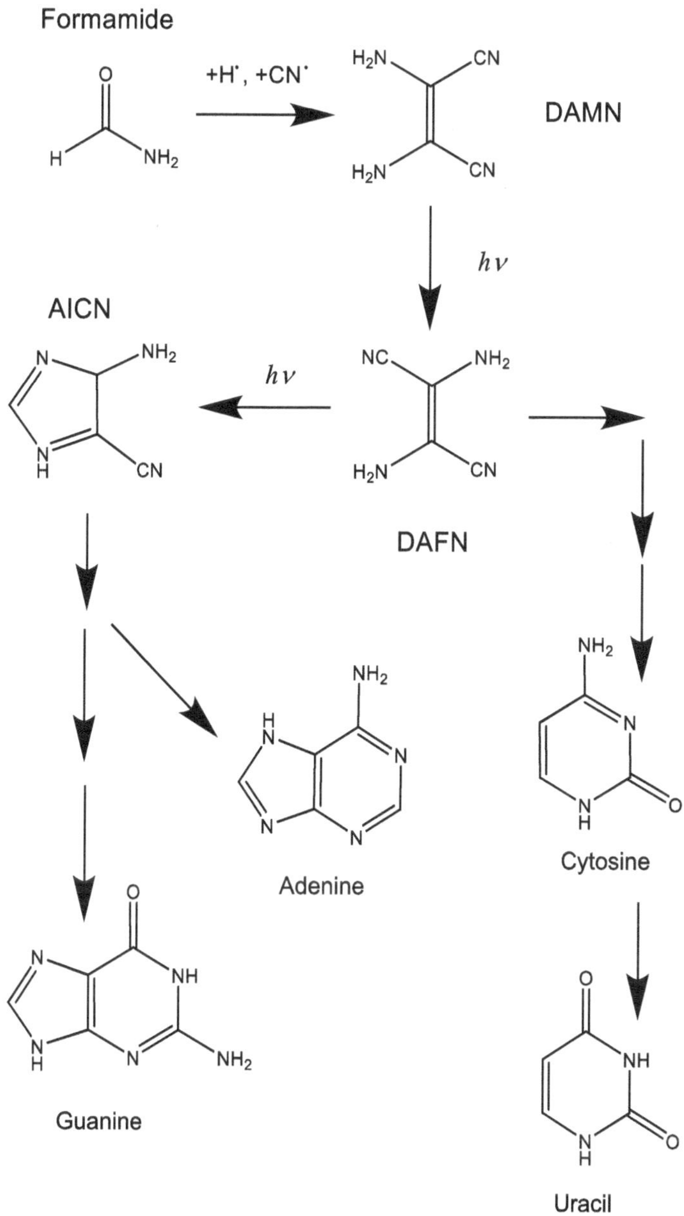

Fig. 24.6.1 Flow of high-energy synthesis from formamide to four canonical nucleobases after Ferus et al. (2015).

Except for cytosine, all canonical nucleobases were detected in the samples of formamide exposed to LIDB. Irradiation of formamide and DAMN using a high-power laser produced adenine, guanine, and uracil, whereas DAMN suspensions produced all of the bases.

When formamide was irradiated in the presence of clay, all four bases were detected. The role of clay is to protect the adsorbed cytosine against deamination to uracil in a further reaction step.

To demonstrate this, formamide was irradiated in the presence of an olivine chondrite meteorite, Northwest Africa 6472 (NWA 6472). Olivine is a silicate mineral with very low or negligible sorption capacity. Indeed, the results of the irradiation with and without chondrite meteorite are roughly similar and clearly show that cytosine was not formed in the presence of olivine, whereas clay supports the formation of all studied bases.

Therefore, the following general model is made for the high-energy synthesis of nucleobases from formamide (Fig. 24.6.1). The synthesis is initiated by a reaction of formamide with the CN• radical forming several intermediates. This part of the reaction pathway leads to DAMN.

The photoisomerization of DAMN produces 2,3-diaminofumaronitrile (DAFN), which binds to another CN• radical to cyclize readily into a tri-substituted pyrimidinyl radical. This moiety serves as the precursor for cytosine and uracil.

Another pathway includes an additional reaction step in which DAFN cyclizes to 4-amino-5-cyanoimidazole (AICN). This synthetic pathway may lead either directly to adenine or the precursor of guanine.

24.7 THE ICY PLANET FALLEN ON THE OCEAN OF THE EARTH

In Section 24.6, we learnt how to build the blocks of life; for example, amino acids and nucleobases or CHONs can be synthesized in ice-ice collisions at high speed. If the CHONs were made at a high rate, and fell to Earth as interplanetary dust particles (IDPs), most origins of life could be extraterrestrial. As discussed in Section 24.6, larger particles can be too hot during entering the Earth, resulting in decomposing, burning, and sterilizing of amino acids, nucleobases, and CHONs.

However, these impacts cause chemical reactions among meteoritic materials, the ocean, and the atmosphere. Formation of reduced volatiles from inorganic materials has been reported in simulations of postimpact reactions on early Earth (Furukawa et al., 2014).

Furthermore, Furukawa's group has investigated such postimpact reactions with experimental simulations and demonstrated the formation of glycine and aliphatic carboxylic acids from inorganic carbon in meteorites. Amino acid formation in impacts involving simulated cometary ice composed of ammonia, methanol, and carbon dioxide has also been proposed. These studies support the importance of impact-induced reactions as a mechanism for providing the building blocks of early life on Earth.

The study of Furukawa's group synthesized glycine using solid amorphous carbon as the carbon source. They presumed high partial pressure of CO_2 in the early Hadean atmosphere, and that therefore dissolution of large quantities of CO_3^{2-} in the early oceans should have occurred. Hence, large amounts of carbon would have been available in the postimpact plumes. The huge carbon reservoir on the early Earth might have been used in impact-induced reactions to form various kinds of organic compounds important for life.

Ammonia can be formed through the reduction of terrestrial nitrogen species in the ocean, crust, and impact plumes (Furukawa et al., 2014). Thus, both CO_3^{2-} and ammonia were easily available in the impact environments, and could have been early carbon and nitrogen sources for prebiotic synthesis on Earth.

The purpose of this section is to investigate what kinds of nucleobases and amino acids are synthesized from CO_3^{2-} and ammonia in simulating the fall of meteorites, after Furukawa et al. (2015). The sample container was made of low carbon stainless steel (SUS 304 L). Gaseous nitrogen was introduced into the head space of the sample container. The shock experiments were conducted using a single-stage propellant gun.

For the shock-recovery experiment, forsterite (Mg_2SiO_4), metallic iron, magnetite (Fe_3O_4), metallic nickel, ^{13}C-labeled ammonium bicarbonate solution (representative of ocean), and gaseous nitrogen (representative of atmosphere) were mixed to be representative of simplified meteorite components (see Table 24.7.1). The mixtures of IM*x*, OC*x*, and CC*x* ($x = 1$ or 2)

Table 24.7.1 Composition of Starting Materials and Impact Velocity (Furukawa et al., 2015)

Type		IM2	IM1	CC1	CC2	OC1	OC2
Starting materials (mg)	Fe	200	300	50	50	100	100
	Fe_3O_4	0	0	100	100	0	0
	Ni	20	30	15	15	30	30
	Mg_2SiO_4	0	0	200	300	200	200
	NH_4HCO_3	170	170	170	40	170	30
	H_2O	130	130	130	150	130	150
	N_2 (gas)	filled	filled	filled	filled	filled	filled
Impact velocity	$km\,s^{-1}$	0.82	0.86	0.89	0.86	0.86	0.87

Table 24.7.2 Run Products of the Impact Experiments (Furukawa et al., 2015)

Products			**IM2**	**IM1**	**CC1**	**CC2**	**OC1**	**OC2**
Nucleobase	(n mol)	Cytosine	8.8	5.3	tr.	tr.	0.11	0.16
		Urascil	0.096	0.023	BD	BD	BD	BD
Amino acids	(n mol)	Gly	520	2900	350	34	370	55
		Ala	48	210	12	0.21	13	0.59
		Ser	0.3	536	0.2	BD	0.51	BD
		Asp	0.9	1.9	BD	BD	BD	BD
		Glu	BD	0.9	BD	BD	BD	0.3
		Val	BD	0.9	BD	BD	BD	BD
		Ile	BD	tr.	BD	tr.	BD	BD
		Leu	BD	tr.	tr.	BD	BD	BD
		Pro	BD	tr.	BD	BD	BD	BD
		Sar	1.9	140	2	0.13	2.2	0.1
		β-Ala	19	86	13	0.11	11	tr.
		α-ABA	19	86	5.1	BD	4.5	BD
		β-AIBA	2.5	22	0.6	BD	0.6	BD
	Total		610	3500	380	35	400	56
	C conversion rate (%)		0.062	0.35	0.037	0.018	0.039	0.03
Amines	(n mol)	Methlamine	3900	20,000	1700	2500	1000	4200
		Ethylamine	460	1400	100	40	52	81
		Propylamine	34	240	15	2.5	6.2	6.1
		Butylamine	0	37	5.1	BD	1.1	2
	Total		4400	22,000	1800	2500	1100	4300

Notes: BD and tr. represent "below detection limit" and "detected in trace amounts," respectively. The C conversion rate is from NH_4HCO_3 into amino acids and nucleobases (atom %).

represent an iron meteorite, an ordinary chondrite, and a carbonaceous chondrite, respectively.

Ultra-high performance liquid chromatography (UHPLC) coupled with tandem mass spectrometry (MS/MS) was used for the detection of the ^{13}C-labeled nucleobases. The ^{13}C-labeling ensured accurate identification of products with removing contamination. The combination of UHPLC and MS/MS facilitated the identification of the specific products. Run products are summarized in Table 24.7.2.

Various amino acids were detected, including glycine (Gly), alanine (Ala), serine (Ser), aspartic acid (Asp), glutamic acid (Glu), valine (Val), leucine (Leu), isoleucine (Ile), proline (Pro), β-alanine (β-Ala), sarcosine (Sar), α-amino-*n*-butyric acid (α-ABA), and β-aminoisobutyric acid (β-AIBA), all of which were ^{13}C-labeled. The experimental results were the first observation that simultaneous formations of various amino acids and nucleobases were synthesized in the single shock experiment.

β-Ala, Sar, α-ABA, and β-AIBA are nonproteinogenic amino acids (see Fig. 24.5.2); therefore, this further excluded the contamination. Gly was produced in the largest amount. Only Asp, Glu, and Val were produced in the IM1 experiment. The ^{13}C-labeled primary amines (methylamine, ethylamine, propylamine, and butylamine) decreased as the length of the alkyl chain increased.

Uracil was formed in the IM compositions only, while cytosine was formed in all experiments. When the yields of the amino acids were normalized to the initial amount of $NH_4H^{13}CO_3$, the yields of amino acids decreased in the order of IM1, IM2, OC1, and CC1, depending on the amounts of metallic iron and nickel in the starting materials. The yields also depended on the initial amounts of NH_4C. The conversion rates were $3.5 \times 10^{-1} - 1.8 \times 10^{-2}$ mol%, which were far higher than those using solid amorphous carbon. This meant that the molecular carbon ($H^{13}CO_3^-$) was more reactive than solid carbon.

In this experiment, the impact velocities were lower (~0.9 km s^{-1}) than the typical impact velocities of large extraterrestrial materials (~20 km s^{-1}), which are limited simply by resistance of the container against the speed. The container burst at higher-velocity impacts. It is estimated that the higher velocity would cause higher temperature, resulting in higher production of organic compounds by carbonate reduction reactions. It should be noted that the labile molecules as nucleobases and amino acids were not formed at peak temperatures; instead, they were formed in postimpact conditions with lower temperature-pressure like this experiment.

As nucleobases and amino acids are generated from a carbon reservoir, such organics can be formed in a CO_2-rich atmosphere. Ammonia could be formed from nitrogen oxides in an N_2-atmosphere by lightning and meteorite impacts, and subsequent reduction by ferrous iron in the ocean or iron sulfides from hydrothermal vents.

It is interesting that, in contrast to this study, carbonaceous chondrites contain more purine than pyrimidine (Callahan et al., 2011). In other words, pyrimidine bases were preferentially formed in this study.

24.8 SUMMARY FOR THE FORMATION OF CHOs AND CHONs

Ferus et al. (2015) proposed that formation of the prebiotic nucleobases through pure formamide ($HCONH_2$) during impacts is the key reaction. However, Miyakawa et al. (2002) showed that formamide accumulation in oceans would not be high enough to facilitate the nucleobase synthesis. The formamide-rich lakes are one solution; however, the land areas appeared to be small according to geological evidence.

The continuous formation of CHONs from atmospheric CO_2 and N_2 in the early ocean environment, and impact-induced reactions using terrestrial carbon sources, seemed possible. However, the experiment of Meinert et al. (2016) has shown the importance of CHOs' syntheses on ices in space again. Therefore, the most feasible origin of CHOs and CHONs on early Earth should be the hybrid of space (ice-ice collision in space) and terrestrial (impact of ice on Earth, especially on the ocean) synthetic sources of CHOs and CHONs.

24.9 THE ORIGIN OF LIFE IN HYDROTHERMAL ENVIRONMENTS

24.9.1 Importance of Hydrothermal Micropore Flows on the Sea Floor

Enrichment of chemical precursors before initiating polymerization of long-chain macromolecules are required to support functions in living systems (e.g., polypeptides, proteins, and nucleic acids). Identifying such suitable mechanisms has become an unsolved question regarding the origin of life. The hydrothermal microenvironments on the sea floor are the only place where active mineral surfaces with disequilibria caused by thermal and chemical gradients exist, because such biochemical complexity is needed for the emergence of life.

The recent finding Russell et al. (2014) of alkaline vent systems such as the Lost City vent in the mid-Atlantic ridge caused great excitement, because geochemical serpentinization gave the surroundings abundant hydrogen at moderate temperatures (150–200°C), and the excess hydrogen can reduce carbon dioxide into methane. Therefore, the alkaline vents could be where the prebiotic chemical precursors necessary to the origin of life are produced (Russell et al., 2014).

24.9.2 Thermal Convection Simulations and Experiments Under the Microscale Hydrothermal Environments

Priye et al. (2017) applied simulations and in situ experiments to understand effects of 3D chaotic thermal convection under microscale hydrothermal environments. The microscale environment accelerated the synchronized mixing of chemical species in the bulk, while simultaneously accelerating enrichment at discrete sidewall locations, so that the microenvironments enhanced the surface reaction kinetics by orders of magnitude.

In the experiments of Priye et al. (2017), a physical mechanism capable of achieving simultaneous mixing and focused enrichment in hydrothermal pore microenvironments was observed. They also found that microscale chaotic advection was established in response to a temperature gradient-promoted bulk homogenization of molecular species and simultaneous transporting species to discrete targeted locations on the bounding sidewalls where they become highly enriched. This process gives a high-order acceleration in surface reaction kinetics under conditions naturally found in subsea hydrothermal microenvironments, suggesting a new way to explain prebiotic emergence of macromolecules from dilute organic precursors, which is an unanswered question regarding the origin of life on Earth.

The study of Priye et al. (2017) showed that chaotic thermal convection gave a mechanism to explain prebiotic emergence of complex biomacromolecules from dilute organic precursors, which is an unanswered question in the origin of life on Earth and in exobiological scenarios such as the Jovian moon Europa or Saturn's moon Enceladus (see Sections 25.3 and 25.4; Hsu et al., 2015).

It should be noted that membrane-like films of oligomers capable of sustaining pH gradients are a precursor of basic metabolic processes (Barge et al., 2014).

REFERENCES

Barge, L.M., Kee, T.P., Doloboff, I.J., Hampton, J.M.P., Ismail, M., et al., 2014. The fuel cell model of abiogenesis: a new approach to origin of life simulations. Astrobiology 14, 254–270.

Callahan, M.P., Smith, K.E., Cleaves, H.J., Ruzicka, J., Stern, J.C., et al., 2011. Carbonaceous meteorites contain a wide range of extraterrestrial nucleobases. Proc. Natl. Acad. Sci. U. S. A. 108, 13995–13998.

Chyba, C., Sagan, C., 1992. Endogenous production, exogenous delivery and impact-shock synthesis of organic molecules: an inventory for the origins of life. Nature 355, 125–132. 1992.

Ferus, M., Michalcikova, R., Shestivska, V., Sponer, J., Sponer, J.E., Civis, S., 2014. High-energy chemistry of formamide: a simpler way for nucleobase formation. J. Phys. Chem. A 118, 719–736.

Ferus, M., Nesvorný, D., Sponer, J., Kubelik, P., Michalcikova, R., Shestivska, V., et al., 2015. High-energy chemistry of formamide: A unified mechanism of nucleobase formation. Proc. Natl. Acad. Sci. U. S. A. 112, 657–662.

Furukawa, Y., Samejima, T., Nakazawa, H., Kakegawa, T., 2014. Experimental investigation of reduced volatile formation by high-temperature interactions among meteorite constituent materials, water, and nitrogen. Icarus 231, 77–82.

Furukawa, Y., Nakazawa, H., Sekine, T., Kobayashi, T., Kakegawa, T., 2015. Nucleobase and amino acid formation through impacts of meteorites on the early ocean. Earth Planet. Sci. Lett. 429, 216–222.
Harada, K., Suzuki, S., 1977. Formation of amino acids from ammonium bicarbonate or ammonium formate by contact glow-discharge electrolysis. Naturwissenschaften 64, 484.
Hsu, H.-W., Postberg, F., Sekine, Y., Shibuya, T., Kempf, S., Horányi, M., et al., 2015. Ongoing hydrothermal activities within Enceladus. Nature 519, 207–210.
Martins, Z., Price, M.C., Goldman, N., Sephton, M.A., Burchell, M.J., 2015. Shock synthesis of amino acids from impacting cometary and icy planet surface analogues. Nat. Geosci. 6, 1045–1049. 2015.
Meinert, C., Myrgorodska, I., de Marcellus, P., Buhse, T., Nahon, L., et al., 2016. Ribose and related sugars from ultraviolet irradiation of interstellar ice analogs. Science 352, 208–212.
Miller, S.L., 1953. A production of amino acids under possible primitive earth conditions. Science 117, 528–529.
Miyakawa, S., Cleaves, H.J., Miller, S.L., 2002. The cold origin of life: a. Implications based on the hydrolytic stabilities of hydrogen cyanide and formamide. Orig. Life Evol. Biosph. 32, 195–208.
Nuevo, M., Meierhenrich, U.J., d'Hendecourt, L., et al., 2007. Enantiomeric separation of complex organic molecules produced from irradiation of interstellar/circumstellar ice analog. Adv. Space Res. 39(400–404).
Oró, J., 1961. Mechanism of synthesis of adenine from hydrogen cyanide under possible primitive earth conditions. Nature 191, 1193–1194.
Priye, A., Yu, Y., Hassan, Y.A., Ugaz, V.M., 2017. Synchronized chaotic targeting and acceleration of surface chemistry in prebiotic hydrothermal microenvironments. PNAS 114, 1275–1280.
Russell, M.J., Barge, L.M., Bhartia, R., Bocanegra, D., Bracher, P.J., Branscomb, E., et al., 2014. The drive to life on wet and icy worlds. Astrobiology 14, 308–343.

FURTHER READING

Ferus, M., Civis, S., Mladek, A., Sponer, J., Juha, L., Sponer, J.E., 2012. On the road from formamide ices to nucleobases: IR-spectroscopic observation of a direct reaction between cyano radicals and formamide in a high-energy impact event. J. Am. Chem. Soc. 134, 20788–20796.
Horka, V., Civis, S., Spirko, V., Kawaguchi, K., 2004. The infrared spectrum of CN in its ground electronic state. Collect Czech Chem Commun 69, 73–89.
Pasek, M., Lauretta, D., 2008. Extraterrestrial flux of potentially prebiotic C, N, and P to the early Earth. Orig. Life Evol. Biosph. 38, 5–21.
Schrenk, M.O., Kelley, D.S., Bolton, S.A., Baross, J.A., 2004. Low archaeal diversity linked to subseafloor geochemical processes at the Lost City Hydrothermal Field, Mid-Atlantic Ridge. Environ. Microbiol. 6, 1086–1095.
Valley, J.W., Peck, W.H., King, E.M., Wilde, S.A., 2002. A cool early Earth. Geology 30, 351–354.

CHAPTER TWENTY FIVE

Water on Other Planets and Dwarf Planets

Contents

25.1 DAWN SPACECRAFT FOUND ORGANICS ON CERES

25.1.1 Dawn Spacecraft and Two Target Asteroids

The Dawn spacecraft was launched by NASA to study asteroids: 4 Vesta and 1 Ceres. Objectives of the Dawn mission were to acquire color images, topographic maps, elemental composition maps, and mineral composition maps, and to measure gravity fields in order to understand their formation, evolution, and roles in forming the terrestrial planets.

First, the Dawn spacecraft went to Vesta, which is rocky, dry, and bright. It has an iron core, mantle, and crust made of basalts. Next, Dawn went to the largest asteroid, Ceres, which is an icy, wet, and dark asteroid with a metal-rich core and clay surface. It is believed that an ocean once existed here, but is now frozen or sublimated. Are there remnants of the ocean or organic materials? Does biological activity remain? These are big questions for Ceres.

https://doi.org/10.1016/B978-0-12-817054-0.00025-4

© 2019 Elsevier Inc.
All rights reserved.

25.1.2 Organic Compounds Were Found on Ceres

In some chondritic meteorites, organic compounds are found. In addition, organics were also detected on cometary nuclei. However, spectral signatures of organics have not been detected on the surfaces of asteroids.

De Sanctis et al. (2016) reported that the bright areas in the Occator crater on Ceres (see Fig. 25.1.1) are consistent with a large amount of sodium carbonate. The carbonates were mixed with a dark component and small amounts of phyllosilicates and ammonium carbonate or ammonium chloride. De Sanctis et al. proposed that these are the solid residue of crystallization of brines and entrained altered solids that reached the surface from below. The heat source may be impact heating or internal temperatures that are above the eutectic temperature of brines.

De Sanctis et al. (2017) subsequently reported that data returned by the Visible and InfraRed Mapping Spectrometer on board the Dawn spacecraft showed a clear detection of an organic absorption feature at 3.4 μm. This signature is characteristic of aliphatic organic matter and is mainly localized on a broad region of ~1000 square km close to ~50 km of the Ernutet crater (see Fig. 25.1.2). The mixed presence on Ceres of ammonia-bearing hydrated minerals, water ice, carbonates, salts, and organic material indicates

Fig. 25.1.1 The surface of Ceres and bright spots in Occator crater. The latter is 92 km in diameter. The intense brightness of these spots may be due to the reflection of sunlight by highly reflective materials on the surface, possibly ice. *(Image credit: NASA/JPL-Caltech/UCLA/MPS/DLR/IDA/PSI: https://www.jpl.nasa.gov/spaceimages/images/largesize/PIA20350_hires.jpg).*

Fig. 25.1.2 Ernutet crater. The magenta color (in the web version) is enhanced to white in the figure. The broken line (in the print version) shows the magenta color area in the original figure. The crater is 52 km in diameter in the northern hemisphere. The magenta areas are associated with evidence of organic material. *(Image credit: NASA/JPL-Caltech/UCLA/MPS/DLR/IDA: https://photojournal.jpl.nasa.gov/jpegMod/PIA21419_modest.jpg).*

a very complex chemical environment, suggesting environments that are favorable to prebiotic chemistry.

25.2 AN ANCIENT LAKE IN GALE CRATER ON MARS

25.2.1 Introduction

Curiosity is the name of the Mars rover, which has been moving inside the ~150-km-diameter Gale impact crater for six years since 2012. The rover's observations confirmed that once there was plenty of water on the surface, making lakes 3.8–3.3 billions years ago. There were streams delivering sediments, which deposited as shown in Fig. 25.2.1. The rocks in the photograph are finely laminated mudstones, which are commonly deposited in a lake (Clavin, 2015).

Before Curiosity, some scientists thought that wind-blown dust and sand were accumulated in dry conditions. However, Curiosity revealed that the wet condition was the correct one. From the map data and Curiosity's

Fig. 25.2.1 A view of Mars by Curiosity. Such views confirmed that water filled this place forming a lake. There were streams delivering sediments (mudstones) which deposited as layers. *(Image credit: NASA/JPL-Caltech/MSSS: http://www.nasa.gov/sites/default/files/thumbnails/image/pia19839-galecrater-main.png).*

camera images, the water depth from the floor was estimated to be 800 m. Since Curiosity landed on Mars, it has investigated Mars' potential as a habitat for past life and the paleoclimate record preserved by sedimentary rocks.

25.2.2 Redox Stratification of the Ancient Lake in Gale Crater

Hurowitz et al. (2017) synthesized geochemical and mineralogical data from lake-bed mudstones collected during the first 1300 martian solar days of rover operations in Gale crater, which hosts a ~5 km tall mountain, Mount Sharp. Hurowitz et al. (2017) present evidence for lake redox stratification, established by depth-dependent variations in atmospheric oxidant and dissolved-solute concentrations. Paleoclimate data preserved in the stratigraphy indicate that the climate changed from colder to warmer conditions. Final geochemical modification by saline fluids is also observed.

25.3 EUROPA

Jupiter's icy moon Europa has a very different surface from that of its rocky neighbor, Io. The Galileo image in Fig. 25.3.1 hints at the possibility of liquid water beneath the icy crust of this moon. The gray white parts of Europa's surface in the right side image of Fig. 25.3.1 are composed of water ice. In contrast, the dark gray regions on the right side of the image may be covered by salts (such as hydrated magnesium-sulfate) and an unknown red component. The dark gray terrain on the left side of the image is caused by some other, unknown contaminant. This global view was obtained in 1997; the finest details that can be discerned are 25 km across (Castillo-Rogez, 2015).

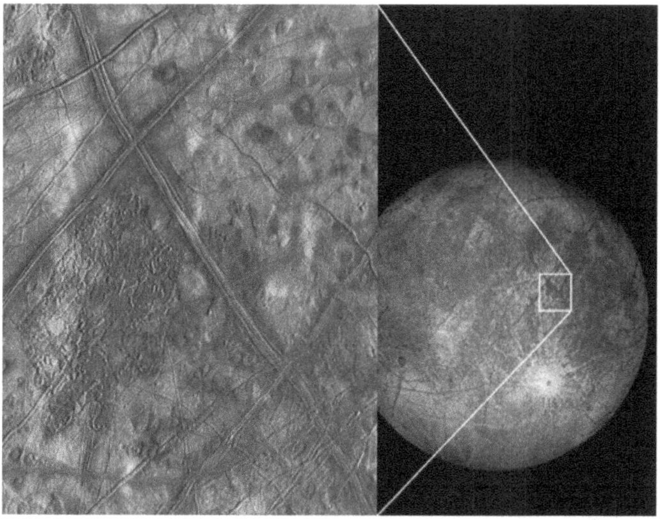

Fig. 25.3.1 Jupiter's icy moon, Europa. This moon has a crust made of blocks, which are thought to have broken apart, as shown in the image on the left. Europa may have had a subsurface ocean at some time. The presence of a magnetic field leads scientists to believe an ocean is present on Europa today. Dark gray areas represent non-ice material resulting from geologic activity. White areas are material ejected during the formation of the impact crater. Long, dark lines are ridges and fractures in the crust. These images were obtained by NASA's Galileo spacecraft. *Image credit: NASA/JPL/University of Arizona.*

Indications of possible plume activity were reported in 2013 by researchers using NASA's Hubble Space Telescope. This telescope was used to observe water vapor above the frigid south polar region of Europa, providing the first strong evidence of water plumes erupting off the moon's surface.

However, NASA's Cassini spacecraft did not find the plume activity during its 2001 flyby of Jupiter. Either the activity was infrequent, or the plumes are smaller than those at Enceladus.

Europa has become one of the most exciting destinations in the solar system for future exploration, because it shows strong indications of having an ocean beneath its icy crust, and thus has a possibility of life.

25.4 ENCELADUS: SATURN'S MOON OF MOVING ICE

Cassini launched in 1997 and entered into an orbit around Saturn in 2004. Enceladus was found to be an icy moon of Saturn. Cassini first discovered the "tiger stripes" on Enceladus, which indicated the young and active surface. Cassini then confirmed that Enceladus has a towering plume of ice (see Fig. 25.4.1). Furthermore, it has water vapor and organic molecules spraying from its south polar region (see Fig. 25.4.1). Cassini determined that the moon has a global ocean and likely hydrothermal activity, meaning it could support simple life (Platt et al., 2014).

The Enceladus plume is thought to come from the ocean below (the blue, dark gray in print versions, region depicted between the core and the white ice layer in Fig. 25.4.1). The spacecraft has flown closer to the surface of Enceladus before, but had never observed such an active plume.

The flyby was not intended to detect life, but it will provide powerful new insights about how habitable the ocean environment is within Enceladus. Cassini scientists are hopeful that the flyby will provide insights about how much hydrothermal activity—that is, chemistry involving rock

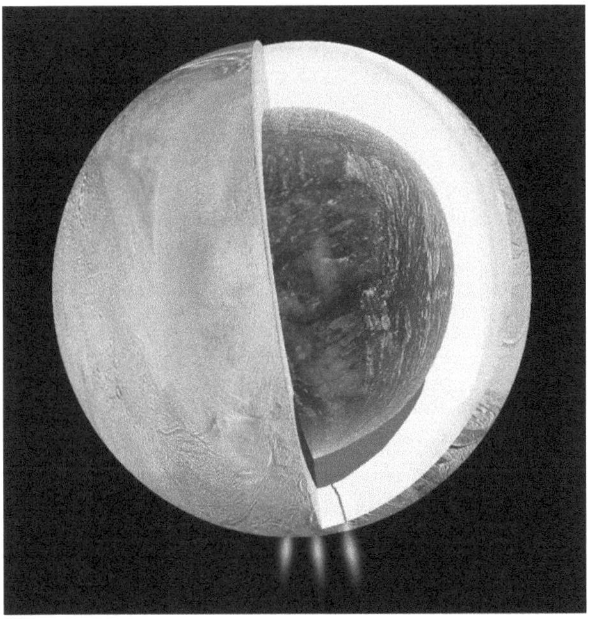

Fig. 25.4.1 The possible interior of Saturn's moon, Enceladus. This diagram illustrates the possible interior of Enceladus based on a gravity investigation by NASA's Cassini spacecraft and NASA's Deep Space Network, reported in 2014. The gravity measurements suggest an ice outer shell and a low density, rocky core with a regional water ocean sandwiched between at high southern latitudes. Views from Cassini's imaging science subsystem were used to depict the surface geology of Enceladus and the plume of water jets gushing from fractures near the moon's south pole. Enceladus is 504 km in diameter. *(Image credit: NASA/JPL-Caltech: http://www.nasa.gov/sites/default/files/pia18071_enceladus-interior.jpeg).*

and hot water—is occurring within Enceladus. This activity could have important implications for the potential habitability of the ocean for simple forms of life. The scientists also expect to understand the chemistry of the plume better as a result.

The flyby will help to solve the mystery of whether the plume is composed of column-like, individual jets, or icy curtain eruptions, or a combination. The answer would make clearer how material is reaching the surface from the ocean below. Researchers are not sure how much icy material the plumes are actually spraying into space. The amount of activity has major implications for how long Enceladus might have been active.

Scientists expected the north polar region of Enceladus to be heavily cratered, based on low-resolution images from the Voyager mission; however, high-resolution Cassini images show a landscape of stark contrast (see Fig. 25.4.2). Thin cracks cross over the pole—the northernmost extent of a global system of such fractures. Before this Cassini flyby, scientists did not know if the fractures extended so far north on Enceladus.

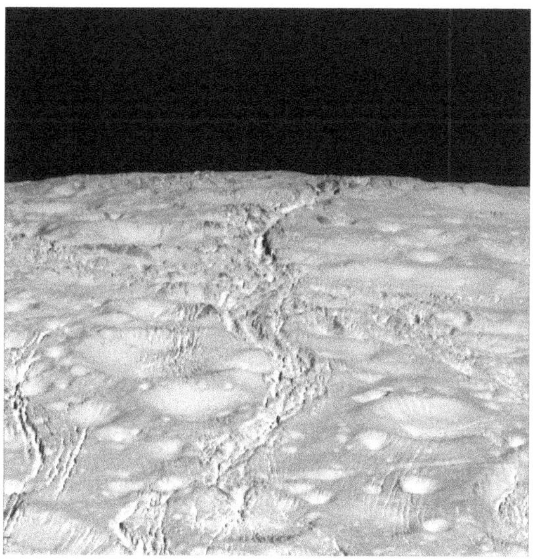

Fig. 25.4.2 Saturn's icy moon, Enceladus. NASA's Cassini spacecraft zoomed Saturn's icy moon Enceladus on October 14, 2015, capturing this image of the moon's north pole. North on Enceladus is up. The view was acquired at a distance of ~6000 km from Enceladus and at a Sun-Enceladus-spacecraft angle of 9 degrees. Image scale is 35 m per pixel. *(Image credit: NASA/JPL-Caltech/Space Science Institute: http://www.nasa.gov/sites/default/files/thumbnails/image/pia19660.jpg).*

REFERENCES

Castillo-Rogez, J., 2015. Solar System Exploration. http://solarsystem.nasa.gov/people/castillo-rogezj.

Clavin, W., 2015. NASA's Curiosity Rover Team Confirms Ancient Lakes on Mars. http://www.nasa.gov/feature/jpl/nasas-curiosity-rover-team-confirms-ancient-lakes-on-mars.

De Sanctis, M.C., Raponi, A., Ammannito, E., Ciarniello, M., Toplis, M.J., et al., 2016. Bright carbonate deposits as evidence of aqueous alteration on (1) Ceres. Nature 536, 54–57.

De Sanctis, M.C., Ammannito, E., McSween, H.Y., Raponi, A., Marchi, S., et al., 2017. Localized aliphatic organic material on the surface of Ceres. Science 355, 719–722.

Hurowitz, J.A., Grotzinger, J.P., Fischer, W.W., McLennan, S.M., Milliken, R.E., et al., 2017. Redox stratification of an ancient lake in Gale crater, Mars. Science 356, 922. eaah6849.

Platt, J., Brown, D., Bell, B., 2014. NASA Space Assets Detect Ocean Inside Saturn Moon. http://solarsystem.nasa.gov/news/2014/04/03/nasa-space-assets-detect-ocean-inside-saturn-moon.

INDEX

Note: Page numbers followed by *f* indicate figures and *t* indicate tables.

D

E

F

I

T

U

FSC
www.fsc.org
MIX
Papier | Fördert
gute Waldnutzung
FSC® C083411

Kolibri 360 GmbH
Ferdinand-Jühlke-Straße 7
99095 Erfurt, Deutschland
produktsicherheit@kolibri360.de